Groundwater Vulnerability Assessment and Mapping Using DRASTIC Model

Groundwater Vulnerability Assessment and Mapping Using DRASTIC Model

Prashant Kumar

Praveen K. Thakur

Sanjit K. Debnath

CRC Press
Taylor & Francis Group
Boca Raton London New York

CRC Press is an imprint of the
Taylor & Francis Group, an **informa** business

CRC Press
Taylor & Francis Group
6000 Broken Sound Parkway NW, Suite 300
Boca Raton, FL 33487-2742

First issued in paperback 2021

© 2020 by Taylor & Francis Group, LLC
CRC Press is an imprint of Taylor & Francis Group, an Informa business

No claim to original U.S. Government works

Printed on acid-free paper

ISBN-13: 978-0-367-25446-9 (hbk)
ISBN-13: 978-1-03-209150-1 (pbk)

Visit the Taylor & Francis Web site at
http://www.taylorandfrancis.com

and the CRC Press Web site at
http://www.crcpress.com

To my wonderful family

To my wonderful parents Raghaw Choubey and Pushpa Choubey; my wife Pinki Choubey—my best friend and constant companion; and Shaurya—my loving son!

Prashant Kumar

To my wonderful family

To my late grandfather, my respected parents Krishan Singh and Saraswati Thakur, my loving wife Jyoti Thakur, and my children Dhruv and Darsh who make my world!

Praveen K. Thakur

To my wonderful family

To my parents

Sanjit K. Debnath

Contents

Preface

There are several books on groundwater vulnerability assessment. So, why this book? It all started during our research related to groundwater vulnerability assessment when we struggled to find a simple, yet robust book full of case studies in the field of groundwater vulnerability assessment. We had to read several research papers to develop the contents for this book to be written in a lucid manner. Several theses are written every year worldwide, but few are read and that forms the so called "Library to Library Research." In this, you choose a research gap from the available literature in the library and solve it and submit the thesis to the library. We are an advocate of "Library to General Mass" research, wherein you solve a problem and bring it to the general people including research scholars, academicians, policymakers, etc., in the form a book. So, this is a book for anyone who wants to work in the field of development of rapid regional assessment tool for the estimation of vulnerability of groundwater to contamination. This book takes Fatehgarh Sahib region (Punjab, India) as a pilot study area and describes how to assess the vulnerability of groundwater to contamination in a very simple manner by stitching the several hydro-geological and human-made factors and highlighting the inherent research problems.

The authors would like to thank all the people who helped make this book a reality. The authors are grateful to Dr. Gagandeep Singh, senior editor and his team members for their tireless efforts and timely feedbacks to make this book in the current state.

Authors

Prashant Kumar is a Scientist in Central Scientific Instruments Organisation, Chandigarh, which is a national laboratory under Council of Scientific and Industrial Research, Government of India. He is also an Honorary Assistant Professor in Academy of Scientific & Innovative Research (AcSIR), Ghaziabad, India. He holds BTech from Manipal Institute of Technology, Manipal, Udupi, India; MTech from AcSIR-CSIO, Chandigarh, India; and PhD in Engineering from AcSIR-CSIO (Central Scientific Instruments Organisation), Chandigarh, India. He has been a Post Graduate Research Programme in Engineering (PGRPE) fellow of CSIR, India. His research interests are agri-informatics, water resource management, agri-instrumentation, and environmental modeling.

 **Praveen K. Thakur** is Scientist/Engineer 'SF' at Water Resources Department (WRD) of Indian Institute of Remote Sensing (IIRS), Indian Space Research Organisation (ISRO), Dehradun, since 2004. He earned his graduate in Civil Engineering from National Institute of Technology (NIT) Hamirpur, post-graduate in Water Resources Engineering from Indian Institute of Technology (IIT) Delhi, and PhD in Geomatics Engineering from IIT Roorkee. His research interests include geospatial technology application in water resources and hydrology; microwave remote sensing for snow, ice, and floods studies; and hydrological and hydraulic modelling. He has published 41 research papers in peer-reviewed international/national journals. Praveen has guided 3 PhD students and more than 25 MTech students. He has participated in Indian Antarctic and Arctic scientific expeditions. He is a life member of six professional societies and recipient of ISRO Young Scientist Merit Award for the year 2014.

Sanjit K. Debnath is a Scientist in Central Scientific Instruments Organisation, Chandigarh under Council of Scientific and Industrial Research. He is also an Honorary Assistant Professor in Academy of Scientific and Innovative Research (AcSIR), Ghaziabad, India. He is a PhD from Indian Institute of Technology, Madras. He has spent a good numbers of years at Korea Advanced Institute of Science and Technology (KAIST), Daejeon, South Korea in the past as a Post-Doctoral Fellow. His research interests are optical instrumentation, digital holography, and remote sensing.

1

Introduction

Groundwater is a valuable resource for the existence of the mankind as people across the globe use it for various activities like consumption, irrigation, and industrial use, etc. Its contamination has always been a big concern for such activities and has aroused curiosity among the researchers, government agencies, and environmental organizations in the recent years. The long term exposure of toxic contaminants present in groundwater has adverse effects on human health in the form of various grievous diseases like skin lesion, skin cancer, neurological effect, hypertension, cardiovascular diseases, pulmonary diseases, and diabetes mellitus (Smith et al. 1992, 2000; Saha et al. 1999; Tseng et al. 2003; Kile and Christiani 2008; Hendryx, 2009). Groundwater contaminants are naturally occurring inorganic pollutants such as arsenic, aluminum, lead, mercury, fluoride, iron, nitrate, etc. and manmade organic pollutants such as pesticides, plasticizers, chlorinated, solvents, etc. (Ghosh and Singh 2009; US Geological Survey 2014) which are widely spread in air, water, soil, rocks, plants, and animals in different ratios. Inorganic contaminants are released in environment via chemical and physical breakdown of rocks and subsequent leaching and runoff, volcanoes, microorganisms, and human activities such as mining and excavation. Organic contaminants are released in environment through various agricultural activities. These contaminants get into groundwater and finally enter into human bodies through food chain. Groundwater is relatively less vulnerable to contamination in comparison to surface water. However, urbanization and industrialization have caused a serious threat to the water resources because the natural purification rate has been subdued by the rate of waste/industrial effluent discharge into the environment. Also, the panoptic use of tube wells has caused serious groundwater intoxication over the years. Many government agencies have come up with their reports on groundwater quality assessment models. Some of the severely affected countries are: Bangladesh (Smith et al. 2000; McArthur et al. 2001;

*This chapter is based on **Prashant Kumar et al.**, Index-based Groundwater Vulnerability Mapping Models using Hydrogeological Settings: A Critical Evaluation, Environmental Impact Assessment Review, Elsevier, Volume 51, February 2015, Pages 38–49 (**Impact Factor-3.09**).*

Hossain et al. 2007; Islam and Islam 2007; Elahi et al. 2012), India (Lalwani et al. 2004; Chakraborty et al. 2007; Ghosh and Singh 2009; Umar et al. 2009; Gorai and Kumar 2013; CGWB 2014), China (Cuihua et al. 2007; Dong et al. 2009; Su et al. 2009; Wang et al. 2010; Cheng et al. 2011; Shuaijun et al. 2011; Wu et al. 2011; Yang et al. 2011; Hamza 2013; Rodríguez-Lado et al. 2013), U.S.A. (USGS 2000; Klug 2009; Ruopu and Lin 2011; USEPA 2014), etc. (Islam and Islam 2007; Rodríguez-Lado et al. 2013; CGWB 2014; USEPA 2014).

In order to monitor and assess the quality of water from the regions where the primary source of drinking water is groundwater wells, several groundwater vulnerability and risk mapping models have been developed. These models estimate the sensitivity of groundwater to contamination, and it is expressed in the form of vulnerability map. The vulnerability map segregates the particular region into several hydrogeological subregions with different levels of severity from contamination point of view (Naqa et al. 2006). There are mainly three kinds of techniques used in the creation of vulnerability assessment maps, viz. statistical techniques (National Research Council 1993; Teso et al. 1996; Troiano et al. 1997; Burkart et al. 1999; Kammoun et al. 2018; Moraru and Hannigan 2018), process-based simulation techniques (Rao et al. 1985; Jury and Ghodrati 1987; Wu and Babcock 1999; Pineros Garcet et al. 2006; Tiktak et al. 2006; Epting et al. 2018), and index-based techniques (Aller et al. 1987; Foster 1987; Stempvoort et al. 1993; Civita 1994; Robins et al. 1994; Vrba and Zaporozec 1994; Doerfliger et al. 1999; Daly et al. 2002; Margane 2003; Jang et al. 2017; Jarray et al. 2017; Neshat and Pradhan 2017; Aslam et al. 2018; Douglas et al. 2018; Hamamin and Nadiri 2018; Mondal et al. 2018; Oroji and Karimi 2018) as shown in Figure 1.1. Statistical techniques find the mapping between the spatial variables and the presence of contaminants in the groundwater. Statistical techniques are not generic in nature as they are mostly used in the assessment of groundwater where similar contaminants are present. Process-based techniques employ simulation models to forecast pollutant movement in groundwater. However, they have shortcomings in the form of unavailability of adequate data, but they are more elaborated than simple index-based techniques. Index-based techniques have the advantages over the rest of two as it resolves their limitations. Index-based techniques are not encumbered by computational complexities and data shortage. This is the reason that index-based techniques are the most preferred for the groundwater vulnerability assessment.

The aim of the current chapter is to review and evaluate the recent progress made in the area of vulnerability assessment of groundwater contamination in the form of discussion of various index-based vulnerability assessment models. This chapter discusses the pros and cons of various index-based vulnerability assessment models and their applications in general. It has also been tried to highlight the underlying research gaps in the present state of knowledge regarding the development of groundwater vulnerability assessment models for further research work.

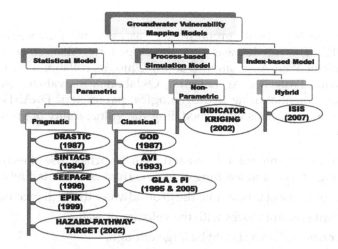

FIGURE 1.1
Index-based groundwater vulnerability mapping models.

Index-Based Vulnerability Mapping Models

Aquifer vulnerability has been modeled in many ways. Of all the techniques developed so far, index-based techniques remain the most widely used techniques because of its large scale aquifer sensitivity and simple implementation. Index-based models can be divided into three categories, namely, parametric, non-parametric, and hybrid models as shown in Figure 1.1.

The following part of this section describes the various index-based models used in the groundwater vulnerability mapping.

Parametric Models

These models are generally represented using a finite number of parameters. Each parameter has its own significance and a particular range split into various intervals. These intervals are assigned a particular value corresponding to the relative sensitivity of the contaminant.

Pragmatic

Pragmatic models are based on observations. They consider all the three important factors in the estimation of zones vulnerable to contamination, namely, the soil conditions, the unsaturated zone of subsoil and bedrock, and the transportation in the saturated zone leading to a final vulnerability index value signifying the extent of contamination (Gogu and Dassargues 2000).

DRASTIC

The most widely used groundwater assessment parametric model is DRASTIC. This model was first developed by Aller et al. (1987) at National Water Well Association Dublin, Ohio, in collaboration with U.S. Environmental Protection Agency Ada, Oklahoma, to evaluate groundwater pollution potential using hydrogeological parameters. DRASTIC model assumes following points while modeling the vulnerability of groundwater contamination:

1. The contaminants are released at the earth surface (use of fertilizers, burning of coal, and leaching of metals from coal-ash tailings)
2. The contaminant flushes into the groundwater through precipitation
3. The contaminant moves with the velocity of water
4. The concerned area should be large enough.

There are two important aspects of this model: hydrogeological parameters and relative ranking system. The model uses seven hydrogeological parameters to assess the groundwater vulnerability, namely, depth to groundwater table (D), net recharge (R), aquifer media (A), soil media (S), topography (T), impact of vadose zone (I), and hydraulic conductivity (C) of the aquifer. These parameters are the most important parameters for groundwater as these are the parameters which regulate the movement of groundwater in a particular study area. These parameters are assigned weights and relative ratings according to their phylogenetic relations to the contaminants. The range of assigned weights is from 1 (least important) to 5 (most important) and relative ratings range is from 1 (least important) to 10 (most important) (Aller et al. 1987; Babiker et al. 2005; Chakraborty et al. 2007). The weights to various parameters are fixed in the implementation of DRASTIC (Aller et al. 1987) as given in Table 1.1. The relative ratings assignment is totally subjective in nature and is based on Delphi consensus (Aller et al. 1987). Depending upon the type of geophysical properties, the parameters are assigned ratings as given in Table 1.1.

DRASTIC model uses a vulnerability index value in order to classify a particular region into different units of potential contamination. The governing equation for the computation of vulnerability index value is given in Equation (1.1).

$$\text{Vulnerability index} = \sum_{i=1}^{7} W_i \times R_i, \tag{1.1}$$

where W_i and R_i represent weight factor and relative rating assigned to the *i*th parameter of DRASTIC vulnerability index empirical formula.

Although DRASTIC considers seven parameters only because of their simple basis to assess the groundwater, nevertheless, this list is not inclusive.

TABLE 1.1

DRASTIC Parameters and Relative Ratings and Weights Assignment

Parameter	Definition	Assigned Weights	Parameter Value/Types	Relative Ratings
Depth to water	It is the actual depth from the earth surface to the water table. It works as a resistive force for the contaminant to sneak through to the groundwater	5	0–5 (feet)	10
			5–15 (feet)	9
			15–30 (feet)	7
			30–50 (feet)	5
			50–75 (feet)	3
			75–100 (feet)	2
			>100 (feet)	1
Net recharge	It is the amount of water per unit area of land which penetrates the ground surface and reaches the aquifer	4	0–2 (inches)	1
			2–4 (inches)	3
			4–7 (inches)	6
			7–10 (inches)	8
			>10 (inches)	9
Aquifer media	It is the porous rock below the earth surface which holds underground water. The type of rock significantly affects the flow of contaminant in the groundwater	3	Massive Shale	1–3
			Metamorphic/Igneous	2–5
			Weathered Metamorphic/ Igneous/Thin Bedded Sandstone, Limestone	3–5
			Shale Sequences	5–9
			Massive Sandstone	4–9
			Massive Limestone	4–9
			Sand and Gravel	4–9
			Basalt	2–10
			Karst Limestone	9–10
Soil media	It is the layer of earth rocks lying between the earth surface and the uppermost bedrock. The percolation of contaminant is highly affected by this layer	2	Thin or Absent	10
			Gravel	10
			Sand	9
			Peat	8
			Shrinking or Aggregated Clay	7
			Sandy Loam	6
			Loam	5
			Silty Loam	4
			Clay Loam	3
			Muck	2
			Non-Shrinking or Non-Aggregated Clay	1

(Continued)

TABLE 1.1 (*Continued*)

DRASTIC Parameters and Relative Ratings and Weights Assignment

Parameter	Definition	Assigned Weights	Parameter Value/Types	Relative Ratings
Topography	It is the gradient of the earth surface and water table follows the similar gradient in the form of conformation line	1	0–2 (percent slope)	10
			2–6 (percent slope)	9
			6–12 (percent slope)	5
			12–18 (percent slope)	3
			>18 (percent slope)	1
Impact of vadose zone	It is the upper portion of the water table in the form of surface soil as well as bedrock layer used for accommodating water. Its impact is measured in terms of porosity, thickness, and permeability	5	Silt/Clay	1–2
			Shale	2–5
			Limestone	2–7
			Sandstone	4–8
			Bedded Limestone, Sandstone, Shale	4–8
			Sand and Gravel with significant silt and clay	4–8
			Metamorphic/Igneous	2–8
			Sand and Gravel	6–9
			Basalt	2–10
			Karst Limestone	6–10
Hydraulic conductivity	It is the rate at which water is transmitted by the bedrock layers. The more the rates are, the higher is the contamination	3	1–100 (GPD/FT^2)	1
			100–300 (GPD/FT^2)	2
			300–700 (GPD/FT^2)	4
			700–1000 (GPD/FT^2)	6
			1000–2000 (GPD/FT^2)	8
			>2000 (GPD/FT^2)	10

GPD: Gallons per day

The implementation of DRASTIC model has been done in association with Geographic information system (GIS) to study several meteorological conditions (Bhaskaran et al. 2001; Chakraborty et al. 2007; Afonso et al. 2008; Dong et al. 2009; Boughriba et al. 2010; Jin et al. 2011; Peter and Sreedevi 2012; Sener and Davraz 2013). It helps in ranking regions of potential vulnerability with respect to various contaminants (Napolitano 1995). The seven parameters of DRASTIC model are shown in Figure 1.2 and are defined in Table 1.1 with their assigned weights and relative ratings according to Aller et al. (1987).

SINTACS

This model was proposed by Civita in 1994. It uses the same parameters as in the case of DRASTIC, but with a different nomenclature (Civita 1994; Civita and De Regibus 1995; Ricchetti and Polemio 2001). SINTACS stands for depth to water (S), net infiltration (I), unsaturated zone (N), soil media (T), aquifer media (A), hydraulic conductivity (C), and slope (S) (Kumar et al. 2012).

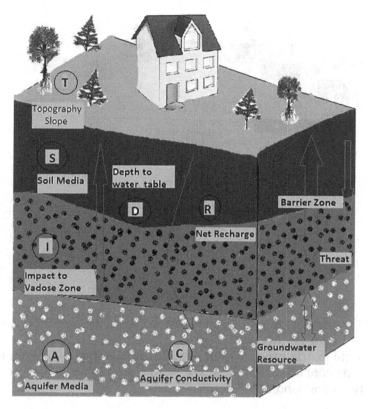

FIGURE 1.2
Definition of DRASTIC parameters.

The difference lies in the way these parameters are assigned weights and relative ratings. The weights are assigned in a more comprehensive manner in order to consider all the environmental conditions related to the seven parameters used in the model. The model employs more than one string of weight assignments to consider the land use factor. This model is of more use for the region where land is used extensively such as coal fields and oil saturated areas. The model calculates the vulnerability indices for different zones. The higher the index value, the higher is the vulnerability. The model is pictorially represented in Figure 1.3. The governing equation for the model is as follows:

$$\text{Vulnerability index} = \sum_{k=1}^{n}\sum_{\substack{i=1 \\ j=1}}^{7}\sum^{7} W_j^k \times R_i X, \tag{1.2}$$

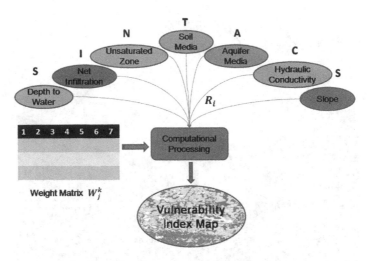

FIGURE 1.3
SINTACS model.

where W_j^k represents the jth weight assigned to ith rating of the particular parameter for the kth weight strings.

The weight matrix plays a significant role in generation of high vulnerable zones in the contaminated areas. The results are highly influenced by the aquifer types and land use patterns.

SEEPAGE

The acronym SEEPAGE stands for system for early evaluation of pollution potential of agricultural groundwater environments (Navulur 1996). This model has been used extensively for estimation of groundwater vulnerability to nitrate contamination on a territorial magnitude (Navulur 1996; Muhammetoglu et al. 2002). Hydrogeological settings and physical properties of the soil are the basis for this model. The model takes into account the following parameters: depth to water table, soil topography, soil depth, impact of vadose zone, aquifer material characteristics, and attenuation potential (Richert et al. 1992). Attenuation potential is a very influential parameter here as it encompasses several significant factors such as soil texture, soil pH, soil permeability, and organic content of the surface. The model uses linear empirical formula similar to DRASTIC model to compute the vulnerability index. SEEPAGE model considers soil properties in more detail than DRASTIC (Richert et al. 1992). However, the range values for the ratings and weights assignments are relatively larger in SEEPAGE model. For example, depending upon the peculiarities of the study area in terms of concentrated or dispersed surfaces, ratings and weights range can be large up to 50; where 50 represents the most significant parameter affecting the groundwater quality. This model also estimates the vulnerability indices for different zones corresponding to a particular empirical formula. The governing equation for the model is as shown below:

$$V_{index} = R_1 \times D_W + R_2 \times S_T + R_3 \times S_D + R_4 \times V_I + R_5 \times A_M + R_6 \times A_P, \quad (1.3)$$

where D_W represents weight assigned to depth of water table; S_T represents weight assigned to soil topography; S_D represents weight assigned to soil depth; V_I represents weight assigned to impact of vadose zone; A_M represents weight assigned to aquifer material; A_P represents weight assigned to attenuation potential; and R_i (i = 1 to 6) represents relative rating assigned to various parameters.

Attenuation Potential is given below:

$$A_P = \sum_{i=1}^{n} \text{Soil parameters } i \times R_i, \quad (1.4)$$

where soil parameters are soil texture, soil pH, soil permeability, etc. R_i (i = 1 to n) represents relative rating assigned to various soil parameters.

EPIK

This model was developed by Doerfliger et al. (1999). It is a multi-dimensional model used for groundwater vulnerability mapping of karst aquifers in particular. The peculiarities of karst aquifers are the meshes of chemical dissolution-based passages and spelunks which allow the speedy and turbulent flow of water. These aquifers are highly vulnerable to arsenic contamination though thin soil or sinks. Even the contaminants disperse very rapidly inside the aquifer due to turbulent flow of water (Margane 2003). Natural filtration of arsenic is relatively not very effective in case of karst aquifers. The model is shown in Figure 1.4. This model considers the following elements:

1. *Epikarst development*
2. *Protective cover potency*
3. *Infiltration stipulation*
4. *Karst mesh growth.*

The model calculates the vulnerability index values according to Equation (1.5) and segregates the regions into different levels of vulnerability, viz. low, moderate, and high.

$$\text{Vulnerability index} = 3 \times E_r + P_r + 3 \times I_r + 2 \times K_r, \quad (1.5)$$

where E_r stands for relative rating assigned to EPIkarst parameter; P_r represents relative rating assigned to protection cover parameter; I_r represents

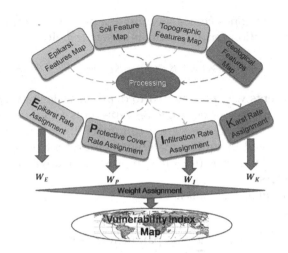

FIGURE 1.4
EPIK model.

relative rating assigned to infiltration parameter; and K_r represents relative rating assigned to karst mesh parameter.

Hazard-Pathway-Target

This model was developed in Europe in order to accomplish consistency in the validation of groundwater intrinsic vulnerability maps. It takes into account all the European conditions for the mapping, and at the same time, it is not rigid thereby giving the liberty to adjust the model according to a particular karstic zones (Daly et al. 2002; Zwahlen 2004). Although this model is primarily focused on karst aquifers, it is capable of considering all kinds of aquifers. The model is based on three keywords: hazards (source/origin), pathway, and target. Hazards (contaminants) are released on ground surface. The target is groundwater resource. And pathway is everything between ground surface and groundwater resource. There are precisely four parameters which are accounted for in this model: precipitation regime, flow concentration, overlaying layers, and karst networks. Precipitation regime (P_{score}) is responsible for total quantity and frequency of water flow-rates. The flow concentration (C_{score}) and overlaying layers (O_{score}) together explain the relative sheltering of groundwater from contaminants. The overlaying layers are made of topsoil, subsoil, non-karstic bedrock, and unsaturated karstic bedrock (Daly et al. 2002; Abdullahi 2009). The method derived from this model is known as COP (Daly et al. 2002; Zwahlen 2004). A factor corresponding to karstic network is introduced to cater the karstic aquifers. The model has been described pictorially in Figure 1.5. The governing equation for the estimation of vulnerability index for the model is as given below.

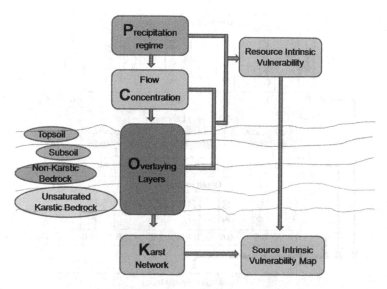

FIGURE 1.5
Hazard-pathway-target model.

$$V_{index} = P_{score} \times C_{score} \times O_{score}. \tag{1.6}$$

Classical

Classical models take into account only the soil conditions and the unsaturated zone. These kinds of assessment techniques ignore the transport operations in the saturated zone (Gogu and Dassargues 2000).

GOD

This is an easy and quick assessment method to map the groundwater vulnerability for contamination as the classical models assume some generic contaminants. It is well suited for indisposed karstification carbonate regions (Shirazi et al. 2012). This model has relatively lesser parameters in comparison to pragmatic models such as DRASTIC, SEEPAGE, SINTACS, and so on. This model depends on three parameters: the groundwater occurrence, overall lithology of the aquifer, and depth to groundwater table (Foster 1987; Robins et al. 1994; Gogu and Dassargues 2000; Shirazi et al. 2012). It doesn't consider the heterogeneities in the used parameters. The model is described pictorially in Figure 1.6. The governing equation of the calculation of vulnerability index for GOD model is given as below:

$$V_{index} = G_r \times O_r \times D_r, \tag{1.7}$$

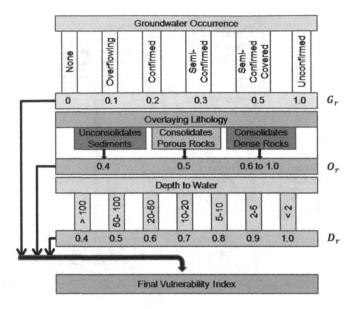

FIGURE 1.6
GOD model.

where G_r is the rating assigned to the groundwater occurrence parameter; O_r is the rating assigned to the overlying lithology parameter; and D_r is the rating assigned to the depth to water table parameter. All these three parameters take values from 0 to 1.

AVI

This method was developed by Stempvoort et al. (1992) in Canada. It is based on two primal elements: measured density in terms of thickness (T) of sedimentary deposits (say *n* deposits) above the topmost aquifer and approximate hydraulic conductivity (C) of each of these deposits (Stempvoort et al. 1993; Vías et al. 2005). In fact, this method considers all the parameters of DRASTIC model except topography and aquifer media ignoring the transportation of contaminants. This model doesn't calculate vulnerability index. Rather, aquifer vulnerability index (AVI) model calculates a theoretical factor called hydraulic resistance (Kruseman and Ridder 1990) for each deposit above the uppermost aquifer as given below:

$$\text{Hydraulic resistance HR} = \sum_{j=1}^{n} T_i / C_i, \qquad (1.8)$$

where T_i and C_i stands for thickness and hydraulic conductivity, respectively, of *i*th deposit layer.

TABLE 1.2

Mapping of Hydraulic Resistance into
Contamination Vulnerability

HR Value	Vulnerability
0–10	Extremely high
10–100	High
100–1000	Moderate
1000–10,000	Low
Greater than 10,000	Extremely low

It's important to note that HR does not represent duration for which contaminants flow. Rather, it signals the travel time of water to move downwards through the poriferous media above the aquifer surface by a phenomenon called advection involving temperature change. Based on the values of HR, AVI model maps the contaminated zone into its severity. AVI model uses Table 1.2 for mapping HR values into contamination vulnerability (Stempvoort et al. 1993).

AVI model is well suited for masking sites for land usage excerption. Limitation of AVI model is that it does not consider separation of aquifers and aquifer water quality (Stempvoort et al. 1993).

GLA & PI

GLA model is a German model (Geologische Landesamter), and it was developed by Hoelting et al. (1995) and Federal Institute for Geosciences and Natural Resources BGR, Hannover. It considers the protective strength of the soil cover and the unsaturated zone against the contaminants such as arsenic and nitrate (Margane 2003). The model considers only the unsaturated zones for vulnerability mapping. In order to evaluate overall protective strength, arguments described in Table 1.3 are taken into assessment.

The governing equation for calculation of protective strength is as follows:

$$\text{Protective strength} = W \times \left(S + \sum_{i=1}^{n} R_i \times T_i \right) + Q + \text{HP}. \tag{1.9}$$

Protective strength defines to what extent is the concerned zone robust enough against the contamination leading to the vulnerability map. GLA model has been successfully tested in many countries such as Jordan, Syria, Lebanon, and South Amman (Margane 2003) over the years and the results have been quite promising.

GLA model is not suitable for karst aquifers as it doesn't consider the discriminatory infiltration paths. In order to make it a generic model, Goldscheider modified GLA model and came out with PI model (Goldscheider 2005) by

TABLE 1.3

Arguments for GLA Model

Arguments	Definition
S	Field capacity of the soil
W	Infiltration rate
R	Rock type
T	Thickness of soil and rock cover above the aquifer
Q	Value to consider perched aquifer systems
HP	Value to consider hydraulic pressure conditions

incorporating two factors: protective cover (P) and infiltration factors (I). PI model is influenced by EPIK model. PI model evaluates the vulnerability of groundwater contamination in terms of spatial distribution of protective strength similar to GLA model represented by π (Vrba and Zaporozec 1994; Tziritis and Evelpidou 2011). The protective strength is calculated as follows:

$$\text{Protective strength } \pi = \sum_{i=1}^{n} P_i \times I_i. \tag{1.10}$$

The value of π varies from 0.0 (lower protection) to 5.0 (higher protection).

Non-parametric Models

Non-parametric models do not make any kind of prejudice regarding the distribution of the data. It makes fewer assumptions and gives due considerations to the data uncertainty while making the predictions of certain data for an unsampled location. Indicator kriging is one such method which is used to locate high potential zones of contamination (Narany et al. 2013).

Indicator Kriging

There are many geostatistical spatial techniques which can be employed for the identification of contaminant deposition over space and time. Kriging is one such non-parametric interpolation technique used for mineral deposit modelling and identification of groundwater contamination severity (Lin et al. 2002; Delbari 2014). It builds probabilistic models of uncertainity relating to the variables chosen for contamination vulnerability mapping (El Alfy 2012; Jang 2013; Narany et al. 2013; Chica-Olmo et al. 2014; Sheikhy et al. 2014). This model is generally employed with DRASTIC model in order to locate optimal sampling points for the study area as kriging alone is not adequate enough for the optimal conception of monitoring wells (Rohazaini et al. 2011). So far, kriging techniqe has been extensively used for the assessment of nitrate contamination in under groundwater in Iran (Narany et al. 2013), Italy (Piccini et al. 2012), and Spain (Caridad et al. 2005).

Hybrid Models

Hybrid models are based on the comparative studies of the various aquifer vulnerability assessment hydrogeological techniques. In general, it uses the architecture of classical models, but imposes the rating and weighting systems of pragmatic methods.

ISIS

ISIS model is a hybrid of pragmatic and classical techniques. It is a river modeling software package. The model is inspired by DRASTIC & SINTACS (pragmatic models) and GOD (classical model) (Civita and De Regibus 1995; Gogu and Dassargues 2000; Macdonald et al. 2007). The parameters considered in this model are the net recharge, topography, soil type and its thickness, lithology and thickness of unsaturated zone, and aquifer medium and its thickness. These parameters are assigned weights and relative ratings according to their phylogenetic relations to the contaminants similar to DRASTIC and SINTACS model. In order to evaluate vulnerability index, following equation is used:

$$V_{\text{index}} = R_I \times W_I + R_S \times W_{SL} \times W_T + R_V \times W_L \times W_{VL} + R_A \times W_{AT} \times W_{AL}, \quad (1.11)$$

where R_I stands for the rating of the net recharge; W_I stands for infiltration coefficient based on land use; R_S stands for the rating for the soil media; W_{SL} stands for soil coefficient based on land use; W_T stands for weighting coefficient based on soil thickness; R_V stands for the rating assigned to the vadose zone; W_L stands for weighting coefficient based on the unsaturated zone lithology and thickness; W_{VL} stands for vadose zone coefficient based on land use; R_A stands for the rating assigned to aquifer media; W_{AT} stands for weighting coefficient based on the aquifer thickness; and W_{AL} stands for aquifer coefficient based on land use.

Comparative Study and Discussion

The modus operandi of all the models has been scientific selection of parameters affecting the groundwater contamination and development of an empirical formula incorporating these parameters which results into vulnerability index values or similar measure such as protective strength or hydraulic resistance for the concerned region. Although there have been attempts to remove the subjectivity associated with the ratings and weights assignment in DRASTIC model, but, at the same time newly evolved models have introduced a different kind of subjectivity such as in the form of assigned ranges of log(c) values to a qualitative aquifer vulnerability index in AVI model (Stempvoort et al. 1993).

The models discussed above have been useful in various geophysical environments. It's very much immanent in nature to say which model is superior over other models. The development of all such models indicate that researchers have worked hard to come up with better results or vulnerability maps over the period of time from 1987 to date. However, each such model has got its limitations along with rewards. And despite advancement in the development of vulnerability mapping models, DRASTIC model still remains the most reliable and accepted model.

There have been quite reasonable efforts in the direction of comparative studies of various models. DRASTIC, SINTACS, GOD, and AVI models have been studied from the perspective of functional qualitative comparison with consideration to diverse families of aquifers like Florina basin in Northern Greece and Piana Campana in Southern Italy (Corniello et al. 1997; Kazakis and Voudouris 2011; Luoma et al. 2017; Aslam et al. 2018; Barzegar et al. 2018; Michalopoulos and Dimitriou 2018). Different models are suitable for different geophysical environmental conditions. No model is generic in nature and can account for all kinds of study regions such as flat, hilly, mountainous, and plain areas of riverside. Table 1.4 presents the characteristics of study area where a particular model is applicable signifying the environmental conditions in the study area.

Groundwater vulnerability assessment technique was started by Aller et al. (1987) and has evolved over the period of time with many modifications such as Modified DRASTIC (Klug 2009), DRTIC (Dong 2009), DRASTIC-Land Pattern (Umar et al. 2009), and DRASTMC-Aquifer Thickness (Yang 2011) in the most widespread model DRASTIC. At the same time, many new models have also been added up. The comparative analysis of various models with its merits and limitations has been presented in Table 1.5.

It is not advisable to compare various vulnerability assessment models because they are using different parameters for their operations, even though all the models have the same objective to assess the vulnerability of groundwater contamination. But, the final vulnerability maps produced by all the vulnerability assessment models do give an insight about the groundwater contamination and its spatial distribution. Such results are validated theoretically by employing two or three models on the same study area and their corresponding vulnerability maps are compared (Vías 2006; Draoui 2008; Ravbar and Goldscheider 2009). But, there is a strong possibility that vulnerability maps produced by two different models might not be exactly same because it all depends on the availability of hydrogeological information corresponding to a particular study area. Therefore, the uncertainty and accuracy assessment remains a challenging task. To get more consistent

TABLE 1.4

Geophysical Conditions and Characteristics of the Study Area for use of Vulnerability Models

Vulnerability Models	Hydrogeological Conditions and Characteristics of Study Area
DRASTIC	Highly populated residential areas, intensive agricultural activities, manufacturing plants and industrial units, arid and semi-arid regions
SINTACS	Intensive mining activities such as coal mines, oil saturated areas, uranium mines, etc., extensive installation of tube wells, carbonated aquifers
EPIK	Karst aquifers, carbonate aquifers
SEEPAGE	Agricultural activities where soil gets affected thereby polluting the groundwater, extreme usage of pesticides and fertilizer
HAZARD-PATHWAY-TARGET	All aquifers
GOD	Poorly karstification carbonate areas like Mediterranean domains, modest variation in vulnerability
AVI	Semi-arid Zones
GLA & PI	Areas with variable lithological units like presence of both karst and non-karst aquifers
INDICATOR KRIGING	Nitrate contaminated areas
ISIS	Intensive land use

results in terms of unification, correlation, and reclassification of the vulnerability indices can be used. At least, this can be validated that the common or overlapping vulnerable declared segments of study area by each of the two or three models are definitely going to be vulnerable. By and large, these maps are validated with clinically analyzed well monitoring data which contains the available contaminants in the samples of different groundwater wells spread across the study area.

Commonly used validation techniques for the vulnerability maps generated by various models are numerical simulations considering hydraulic and transport parameters of study area (Neukum and Hötzl 2007; Neukum 2008) and Bayesian Data Fusion framework (Mattern 2012) hydrographs, chemographs, bacteriological analyses, tracer techniques, water balances, and analogue studies (Goldscheider 2001; Daly et al. 2002; Gogu 2003; Graham and Polizzotto 2013). Since groundwater vulnerability assessment studies rely heavily on availability of hydrogeological data. These data set are correlated with specific properties (depth of water table, hydraulic conductivity, soil type, and so on) of study area to build a predictive model for vulnerability maps (Troiano et al. 1999). The statistical methods used for the validation are correlation coefficient and analysis of co-variance and variance derived from groundwater contaminants (Holman et al. 2005; Sorichetta et al. 2011).

TABLE 1.5

Merits and Limitations of Vulnerability Models

Vulnerability Models	Merits	Limitations
DRASTIC	• Widely accepted model • Economical and less time consuming to evaluate a broader range of groundwater vulnerability • Most suitable for land use management	• Just a qualitative evaluation tool • Land use is a significant factor and rescaling of relative rating and weights need in order to incorporate it • Most subjective model because of flexible ratings assigned to the parameters depending upon the circumstances • Only areas more than 100 acres can be assessed for its vulnerability • Effect of pollution type is not taken into account
SINTACS	• Similar to DRASTIC, but weights are assigned in a more comprehensive manner in order to consider all the environmental conditions related to the seven parameters used in the model	• Complex structure • A number of weight strings are run in parallel
EPIK	• Adequate for karst aquifers • Subjectivity associated is less because more selective choice of variables and reduced relative ratings	• Requirement of elaborated assessment of karst aquifers which is costly and time consuming as it involves thorough studies of geophysics and hydraulic character • Karst features like shallow holes necessitates the interpretation of satellite imagery
SEEPAGE	• Considers the soil parameter in most comprehensive manner	• Larger range for the ratings and weights assignments
HAZARD-PATHWAY-TARGET	• Flexible enough to adjust the model according to a particular karstic zones • Focused on karst aquifers, but it has the capability to consider all kinds of aquifers	• Determination of environmental scenarios and their effects for P with O and C • Under testing in European countries

(Continued)

TABLE 1.5 (*Continued*)

Merits and Limitations of Vulnerability Models

Vulnerability Models	Merits	Limitations
GOD	• Adequate for poorly karstification carbonate area • Takes lesser parameters than DRASTIC so useful if the data availability is small • Influence of rainfall variation is visible on vulnerability maps • Subjectivity associated is less because more selective choice of variables and reduced relative ratings	• Best suited for large areas • Doesn't consider the heterogeneities in the used parameters
AVI	• Numbers of parameters taken is only two • Doesn't consider relative ratings and weights thereby removing the subjectivity issue • Suitable for land use management • Influence of rainfall variation is visible on vulnerability maps	• Not suitable for Karst aquifers • Reports a higher degree of vulnerability of the aquifer in comparison to other models • Many significant parameters have not been taken into account such as climatic conditions, hydraulic gradient, porosity and water content of the porous media, and sorptive or reactive properties of the layers • Doesn't consider aquifer water quality and separation of aquifers • Applicable only for nearest to surface aquifers
GLA	• Most suitable model as scientific considerations are taken for ratings range	• GLA model is not suitable for karst aquifers
PI	• Most suitable model for karstic aquifers • PI model takes into account all kinds of hydrogeological settings	• No consideration to physical attenuation process

(Continued)

TABLE 1.5 (*Continued*)

Merits and Limitations of Vulnerability Models

Vulnerability Models	Merits	Limitations
INDICATOR KRIGING	• Used extensively for the assessment of nitrate contamination in under groundwater • New advancement in vulnerability mapping models	• Not fully established and widespread
ISIS	• Hybrid model with flexibility of having other parameters such as rainfall value and temperature	• Complex in nature

Conclusion and Further Recommendations

No vulnerability assessment model is robust enough which can cater to the needs of all kinds of geological environments. Each study area has its own characteristics. So, one must include the modification factor in his/her study in order to make the model work for his/her study area. In fact, there have been tremendous modifications in all the models as per the needs of the study area and subjected to the restrictions imposed by hydrogeological data availability. The model selection for a particular area depends on many factors such as the scale of mapping, data availability, hydrogeological setting, and the end use of the map. Depending upon the assessment needed, a particular model with suitable modification and inclusive of some other environmental parameters can be employed.

Groundwater vulnerability maps have many advantages. It is very much useful for groundwater quality monitoring and land use planning. These maps assist the government bodies in making the policies related to the planning of land use and water resource management, urban planning, plant or industry establishment, special economic zones, and agricultural activities. Such maps empower the decision makers to contrive an effective strategy for groundwater monitoring.

Over the period of time, hydrogeologists have been resolving many issues related to the development of a groundwater vulnerability assessment model which is quite praiseworthy. However, there are many research gaps related to risk mapping, assessment techniques, and scientific considerations behind inclusion/exclusion of any parameter which are yet to be addressed. The research gaps are as follows:

1. Scientific explanation of relative ratings and weights assigned to various parameters in DRASTIC model because as of now it is based on expert's opinion.
2. 3-D representation of groundwater well data.
3. Very little has been done with regard to the scientific relationship among the various parameters being used by vulnerability models. Further research needs to establish this.
4. Rigorous study needs to be developed for the protection of groundwater wells.
5. Statistical models needs to be developed for the theoretical validation of vulnerability maps generated.
6. Application of various models on diversified meteorological and hydrological settings in order to evaluate the quality of vulnerability maps and its accuracy.
7. Groundwater wells data are full of many contaminants. Study needs to be done to make specific vulnerable maps with respect to a particular contaminant.
8. Development of more hybrid models with ability to adjust more hydrogeological parameters.
9. Intensive mapping needs to be done across the whole world in order to locate the corrupted groundwater wells with a view to protect it further. Bangladesh is one such country where mapping has been done at an extensive level. Recently, India has also shown interest in mass-scale mapping of its various states. USA and European countries have been doing it for quite some time and have well established vulnerable maps.

References

Abdullahi, U. S. (2009). "Evaluation of models for assessing groundwater vulnerability to pollution in Nigeria." *Bayero Journal of Pure and Applied Sciences* 2(2): 5.

Afonso, J. M., A. Pires et al. (2008). "Aquifer vulnerability assessment of urban areas using a gis-based cartography: Paranhos groundwater Pilot Site, Porto, Nw Portugal." *Global Groundwater Resources and Management* 6(14): 20.

Aller, L., T. Bannet et al. (1987). *DRASTIC: A Standardized System to Evaluate Ground Water Pollution Potential Using Hydrogeologic Settings.* Worthington, OH, National Water Well Association.

Aslam, R. A., S. Shrestha et al. (2018). "Groundwater vulnerability to climate change: A review of the assessment methodology." *Science of the Total Environment* **612**: 853–875.

Babiker, I. S., M. A. Mohamed et al. (2005). "A GIS-based DRASTIC model for assessing aquifer vulnerability in Kakamigahara Heights, Gifu Prefecture, central Japan." *Science of the Total Environment* **345**(1–3): 127–140.

Barzegar, R., A. A. Moghaddam et al. (2018). "Mapping groundwater contamination risk of multiple aquifers using multi-model ensemble of machine learning algorithms." *Science of the Total Environment* **621**: 697–712.

Bhaskaran, S., B. Forster et al. (2001). Hail storm vulnerability assessment by using hyperspectral remote sensing and GIS techniques. *Geoscience and Remote Sensing Symposium. IGARSS '01. IEEE 2001 International.*

Boughriba, M., A. Barkaoui et al. (2010). "Groundwater vulnerability and risk mapping of the Angad transboundary aquifer using DRASTIC index method in GIS environment." *Arabian Journal of Geosciences* **3**(2): 207–220.

Burkart, M. R., D. W. Kolpin et al. (1999). "Agrichemicals in ground water of the midwestern USA: Relations to soil characteristics." *Journal of Environmental Quality* **28**(6): 1908–1915.

Caridad Cancela, R., E. V. Vázquez et al. (2005). "Assessing the spatial uncertainty of mapping trace elements in cultivated fields." *Communications in Soil Science and Plant Analysis* **36**(1–3): 253–274.

Central Ground Water Board (2014). "Groundwater contamination." from http://www.cgwb.gov.in/ (accessed 25 April 2014).

Chakraborty, S., K. P. Paul et al. (2007). "Assessing aquifer vulnerability to arsenic pollution using DRASTIC and GIS of North Bengal Plain: A case study of English Bazar Block, Malda District, West Bengal, India." *Journal of Spatial Hydrology* **7**(1): 21.

Cheng, D., J. Jia et al. (2011). Intrinsic vulnerability of groundwater resources in a loess area in northern China. *Water Resource and Environmental Protection (ISWREP), 2011 International Symposium on.* Xi'an, China, IEEE.

Chica-Olmo, M., E. Pardo-Igúzquiza et al. (2014). Quantitative risk management of groundwater contamination by nitrates using indicator geostatistics. *Mathematics of Planet Earth.* E. Pardo-Igúzquiza, C. Guardiola-Albert, J. Heredia et al. (Eds.), Berlin, Germany, Springer, pp. 533–536.

Civita, M. V. (1994). *Le carte della vulnerabilità degli acquiferi all'inquinamento: Teoria & pratica* [Groundwater vulner-ability maps to contamination: Theory and practice]. Bologna, Italy, Pitagora, p. 325.

Civita, M. and C. De Regibus (1995). "Sperimentazione di alcune metodologie per la valutazione della vulnerabilità degli aquiferi." *Quaderni di Geologia Applicata Pitagora* **3**: 63–71.

Corniello, A., D. Ducci et al. (1997). Comparison between parametric methods to evaluate aquifer pollution vulnerability using a GIS: An example in the "Piana Campana," southern Italy. *Engineering Geology and the Environment International Symposium; 22nd, Engineering Geology and the Environment.* Rotterdam, the Netherlands, Ashgate Publishing Company.

Cuihua, C., N. Shijun et al. (2007). Assessing spatial-temporal variation of heavy metals contamination of sediments using GIS 3D spatial analysis methods in Dexing mines, Jiangxi province, China. *Geoscience and Remote Sensing Symposium. IGARSS 2007. IEEE International.*

Daly, D., A. Dassargues et al. (2002). "Main concepts of the 'European approach' to karst-groundwater-vulnerability assessment and mapping." *Hydrogeology Journal* **10**(2): 340–345.

Delbari, M., M. Amiri et al. (2014). "Assessing groundwater quality for irrigation using indicator kriging method." *Applied Water Science* **6**: 371–381.

Doerfliger, N., P. Y. Jeannin et al. (1999). "Water vulnerability assessment in karst environments: A new method of defining protection areas using a multi-attribute approach and GIS tools (EPIK method)." *Environmental Geology* **39**(2): 165–176.

Dong, Z., Y. Wei et al. (2009). Groundwater vulnerability assessment using DRTIC model in Jiaozuo City, China. *Bioinformatics and Biomedical Engineering, 2009. ICBBE 2009. 3rd International Conference on.*

Douglas, S. H., B. Dixon et al. (2018). "Assessing intrinsic and specific vulnerability models ability to indicate groundwater vulnerability to groups of similar pesticides: A comparative study." *Physical Geography* **39**: 1–19.

Draoui, M., J. Vias et al. (2008). "A comparative study of four vulnerability mapping methods in a detritic aquifer under mediterranean climatic conditions." *Environmental Geology* **54**(3): 455–463.

El Alfy, M. (2012). "Integrated geostatistics and GIS techniques for assessing groundwater contamination in Al Arish area, Sinai, Egypt." *Arabian Journal of Geosciences* **5**(2): 197–215.

Elahi, M. M., M. I. Amin et al. (2012). An optimal resource distribution method for arsenic mitigation in Bangladesh. *Computer and Information Technology (ICCIT), 15th International Conference on.* Chittagong, India, IEEE.

Epting, J., R. M. Page et al. (2018). "Process-based monitoring and modeling of Karst springs—Linking intrinsic to specific vulnerability." *Science of the Total Environment* **625**: 403–415.

Foster, S. S. D. (1987). *Fundamental Concepts in Aquifer Vulnerability, Pollution Risk and Protection Strategy.* The Hague, the Netherlands, Netherlands Organization for Applied Scientific Research.

Ghosh, N. C. and R. D. Singh (2009). "Groundwater arsenic contamination in India: Vulnerability and scope for remedy." National Institute of Hydrology, from http://www.researchgate.net/publication/242557859_Groundwater_Arsenic_Contamination_in_India_Vulnerability_and_Scope_for_Remedy (accessed 20 May 2014).

Gogu, R. C. and A. Dassargues (2000). "Current trends and future challenges in groundwater vulnerability assessment using overlay and index methods." *Environmental Geology* **39**(6): 549–559.

Gogu, R., V. Hallet et al. (2003). "Comparison of aquifer vulnerability assessment techniques. Application to the Néblon river basin (Belgium)." *Environmental Geology* **44**(8): 881–892.

Goldscheider, N. (2005). "Karst groundwater vulnerability mapping: Application of a new method in the Swabian Alb, Germany." *Hydrogeology Journal* **13**(4): 555–564.

Goldscheider, N., H. Hötzl et al. (2001). "Validation of a vulnerability map (EPIK) with tracer tests." *Sciences et techniques de l'environnement. Mémoire hors-série* **13**: 167–170.

Gorai, K. A. and S. Kumar (2013). "Spatial distribution analysis of groundwater quality index using GIS: A case study of ranchi municipal corporation (RMC) area." *Geoinformatics & Geostatistics: An Overview* **1**(2): 11.

Graham, J. P. and M. L. Polizzotto (2013). "Pit latrines and their impacts on groundwater quality: A systematic review." *Environmental Health Perspectives* **121**: 10.

Hamamin, D. F. and A. A. Nadiri (2018). "Supervised committee fuzzy logic model to assess groundwater intrinsic vulnerability in multiple aquifer systems." *Arabian Journal of Geosciences* **11**(8): 176.

Hamza, M. (2013). Millions face arsenic contamination risk in China, study finds. *The Guardian*, China.

Hendryx, M. (2009). "Mortality from heart, respiratory, and kidney disease in coal mining areas of Appalachia." *International Archives of Occupational and Environmental Health* **82**(2): 243–249.

Hoelting, B., T. Haertle et al. (1995). *Concept for the Determination of the Protective Effectiveness of the Cover Above the Groundwater Against Pollution.* Hannover, Germany, Ad-hoc Working Group on Hydrogeology, 28 p.

Holman, I. P., R. Palmer et al. (2005). "Validation of an intrinsic groundwater pollution vulnerability methodology using a national nitrate database." *Hydrogeology Journal* **13**(5–6): 665–674.

Hossain, F., J. Hill et al. (2007). "Geostatistically based management of arsenic contaminated ground water in shallow wells of Bangladesh." *Water Resources Management* **21**(7): 1245–1261.

Islam, S. M. and F. Islam (2007). "Arsenic contamination in groundwater in Bangladesh: An environmental and social disaster." from http://www.iwawaterwiki.org/xwiki/bin/view/Articles/Arsenic (accessed 25 May 2014).

Jang, C.-S. (2013). "Use of multivariate indicator kriging methods for assessing groundwater contamination extents for irrigation." *Environmental Monitoring and Assessment* **185**(5): 4049–4061.

Jang, W., B. Engel et al. (2017). "Aquifer vulnerability assessment for sustainable groundwater management using DRASTIC." *Water* **9**(10): 792.

Jarray, H., M. Zammouri et al. (2017). "GIS based DRASTIC model for groundwater vulnerability assessment: Case study of the shallow mio-plio-quaternary aquifer (Southeastern Tunisia)." *Water Resources* **44**(4): 595–603.

Jin, H., H. Liu et al. (2011). Research progress of a GIS-based DRASTIC model. *Mechanic Automation and Control Engineering (MACE), 2011 Second International Conference on.* Hohhot, China, IEEE.

Jury, W. A. and M. Ghodrati (1987). Overview of organic chemical environmental fate and transport modeling approaches. Reactions and movement of organic chemicals in soils. *Proceedings of a Symposium of the Soil Science Society of America and the American Society of Agronomy.* Atlanta, GA, Soil Science Society of America.

Kammoun, S., R. Trabelsi et al. (2018). "Groundwater quality assessment in semiarid regions using integrated approaches: The case of Grombalia aquifer (NE Tunisia)." *Environmental Monitoring and Assessment* **190**(2): 87.

Kazakis, N. and K. Voudouris (2011). Comparison of three applied methods of groundwater vulnerability mapping: A case study from the Florina basin, Northern Greece. *Advances in the Research of Aquatic Environment*. N. Lambrakis, G. Stournaras and K. Katsanou (Eds.), Berlin, Germany, Springer, pp. 359–367.

Kile, M. L. and D. C. Christiani (2008). "Environmental arsenic exposure and diabetes." *Journal of the American Medical Association* **300**(7): 845–846.

Klug, J. (2009). "Modeling the risk of groundwater contamination using DRASTIC and geographic information systems in Houston County, Minnesota." *Papers in Resource Analysis* **11**: 12.

Kruseman, G. P. and N. A. Ridder (1990). *Analysis and Evaluation of Pumping Test Data*. Wageningen, the Netherlands, ILRI Publication.

Kumar, S., D. Thirumalaivasan et al. (2012). "Groundwater vulnerability assessment using SINTACS model." *Geomatics, Natural Hazards and Risk* **4**(4): 339–354.

Lalwani, S., D. T. Dogra et al. (2004). "Study on arsenic level in ground water of delhi using hydride generator accessory coupled with atomic absorption spectrophotometer." *Indian Journal of Clinical Biochemistry* **19**(2): 6.

Lin, Y. P., T. K. Chang et al. (2002). "Factorial and indicator kriging methods using a geographic information system to delineate spatial variation and pollution sources of soil heavy metals." *Environmental Geology* **42**(8): 900–909.

Luoma, S., J. Okkonen et al. (2017). "Comparison of the AVI, modified SINTACS and GALDIT vulnerability methods under future climate-change scenarios for a shallow low-lying coastal aquifer in southern Finland." *Hydrogeology Journal* **25**(1): 203–222.

Macdonald, D., R. Hall et al. (2007). Investigating the interdependencies between surface and groundwater in the Oxford area to help predict the timing and location of groundwater flooding and to optimise flood mitigation measures. *42nd Defra Flood and Coastal Management Conference*. York, UK.

Margane, A. (2003). *Guideline for Groundwater Vulnerability Mapping and Risk Assessment for the Susceptibility of Groundwater Resources to Contamination*, Vol. 4. Damascus, Syria, BGR and ACSAD, p. 177.

Mattern, S., W. Raouafi et al. (2012). "Bayesian data fusion (BDF) of monitoring data with a statistical groundwater contamination model to map groundwater quality at the regional scale." *Journal of Water Resource and Protection* **4**(11): 929–943.

McArthur, J. M., P. Ravenscroft et al. (2001). "Arsenic in groundwater: Testing pollution mechanisms for sedimentary aquifers in Bangladesh." *Water Resources Research* **37**(1): 109–117.

Michalopoulos, D. and E. Dimitriou (2018). "Assessment of pollution risk mapping methods in an Eastern Mediterranean catchment." *Journal of Ecological Engineering* **19**(1): 55–68.

Mondal, N. C., S. Adike et al. (2018). *Assessing Aquifer Vulnerability Using GIS-Based DRASTIC Model Coupling with Hydrochemical Parameters in Hard Rock Area from Southern India*. Singapore, Springer.

Moraru, C. and R. Hannigan (2018). Geochemical method of the groundwater vulnerability assessment. *Analysis of Hydrogeochemical Vulnerability*. Cham, Switzerland, Springer International Publishing, pp. 17–40.

Muhammetoglu, H. and A. Muhammetoglu et al. (2002). "Vulnerability of ground-water to pollution from agricultural diffuse sources: A case study." *Water Science and Technology* **45**(9): 7.

Napolitano, P. (1995). *Assessing Aquifer Vulnerability to Pollution in the Piana Campana.* Enschede, the Netherlands, ITC.

Naqa, E. A., N. Hammouri et al. (2006). "GIS-based evaluation of groundwater vulnerability in the Russeifa area, Jordan." *Revista Mexicana de Ciencias Geologicas* **23**(6): 11.

Narany, T. S., M. Ramli et al. (2014). "Groundwater irrigation quality mapping using geostatistical techniques in Amol–Babol Plain, Iran." *Arabian Journal of Geosciences* **8**: 961–976.

Narany, T. S., M. F. Ramli et al. (2013). "Spatial assessment of groundwater quality monitoring wells using indicator kriging and risk mapping, Amol-Babol Plain, Iran." *Water* **6**(1): 68–85.

National Research Council (1993). *Ground Water Vulnerability Assessment: Contamination Potential Under Conditions of Uncertainty.* Washington, DC, National Academy Press.

Navulur, K. C. S. (1996). Groundwater vulnerability evaluation to nitrate pollution on a regional scale using GIS. Doctoral dissertations & theses, West Lafayette, IN, Purdue University.

Neshat, A. and B. Pradhan (2017). "Evaluation of groundwater vulnerability to pollution using DRASTIC framework and GIS." *Arabian Journal of Geosciences* **10**(22): 501.

Neukum, C. and H. Hötzl (2007). "Standardization of vulnerability maps." *Environmental Geology* **51**(5): 689–694.

Neukum, C., H. Hötzl et al. (2008). "Validation of vulnerability mapping methods by field investigations and numerical modelling." *Hydrogeology Journal* **16**(4): 641–658.

Oroji, B. and Z. F. Karimi (2018). "Application of DRASTIC model and GIS for evaluation of aquifer vulnerability: Case study of Asadabad, Hamadan (western Iran)." *Geosciences Journal* **22**(5): 843–855.

Peter, S. and C. Sreedevi (2012). Water quality assessment and GIS mapping of ground water around KMML industrial area, Chavara. *Green Technologies (ICGT), 2012 International Conference on.* Trivandrum, India, IEEE.

Piccini, C., A. Marchetti et al. (2012). "Application of indicator kriging to evaluate the probability of exceeding nitrate contamination thresholds." *International Journal of Environmental Research* **6**(4): 10.

Pineros Garcet, J. D., A. Ordonez et al. (2006). "Metamodelling: Theory, concepts and application to nitrate leaching modelling." *Ecological Modelling* **193**(3–4): 629–644.

Rao, P. S. C., A. C. Hornsby et al. (1985). Indices for ranking the potential for pesticide contamination in groundwater. *Proceedings of Soil Crop Science Society.* Gainesville, FL, University of Florida.

Ravbar, N. and N. Goldscheider (2009). "Comparative application of four methods of groundwater vulnerability mapping in a Slovene karst catchment." *Hydrogeology Journal* **17**(3): 725–733.

Ricchetti, E. and M. Polemio (2001). Vulnerability mapping of carbonate aquifer using geographic information systems. *Geoscience and Remote Sensing Symposium. IGARSS '01. IEEE 2001 International.*

Richert, S. E., S. E. Young et al. (1992). SEEPAGE: A GIS model for groundwater pollution potential. *ASAE International Winter Meeting*, Nashville, TN.

Robins, N., B. Adams et al. (1994). "Groundwater vulnerability mapping: The British perspective." *Hydrogéologie* 3: 35–42.

Rodríguez-Lado, L., G. Sun et al. (2013). "Groundwater arsenic contamination throughout China." *Science* 341(6148): 866–868.

Rohazaini, M. J., N. S. Mohamad et al. (2011). Geostatistics approach with indicator kriging for assessing groundwater vulnerability to Nitrate contamination. *7th ESRI Asia Pacific User Conference & 21st Korean GIS Conference*, Seoul, Korea, p. 10.

Ruopu, L. and Z. Lin (2011). Vadose zone mapping using geographic information systems and geostatistics a case study in the Elkhorn River Basin, Nebraska, USA. *Water Resource and Environmental Protection (ISWREP), 2011 International Symposium on.* Xi'an, China, IEEE.

Saha, J. C., A. K. Dikshit et al. (1999). "A review of arsenic poisoning and its effects on human health." *Critical Reviews in Environmental Science and Technology* 29(3): 281–313.

Sener, E. and A. Davraz (2013). "Assessment of groundwater vulnerability based on a modified DRASTIC model, GIS and an analytic hierarchy process (AHP) method: The case of Egirdir Lake basin (Isparta, Turkey)." *Hydrogeology Journal* 21(3): 701–714.

Shirazi, S. M., H. M. Imran et al. (2012). "GIS-based DRASTIC method for groundwater vulnerability assessment: A review." *Journal of Risk Research* 15(8): 991–1011.

Shuaijun, H., Z. Maosheng et al. (2011). A GIS-based groundwater vulnerability assessment for the Northern Shaanxi energy and chemical base, China. *Water Resource and Environmental Protection (ISWREP), 2011 International Symposium on.* Xi'an, China, IEEE.

Smith, A. H., C. Hopenhayn Rich et al. (1992). "Cancer risks from arsenic in drinking water." *Environmental Health Perspectives* 97: 9.

Smith, A. H., E. O. Lingas et al. (2000). "Contamination of drinking-water by arsenic in Bangladesh: A public health emergency." *Bulletin of the World Health Organization* 78: 1093–1103.

Sorichetta, A., M. Masetti et al. (2011). "Reliability of groundwater vulnerability maps obtained through statistical methods." *Journal of Environmental Management* 92(4): 1215–1224.

Stempvoort, D. V., L. Ewert et al. (1992). *AVI: A Method for Groundwater Protection Mapping in the Prairie Provinces of Canada.* Regina, Saskatchewan, Prairie Provinces Water Board.

Stempvoort, D. V., L. Ewert et al. (1993). "Aquifer vulnerability index: A gis: Compatible method for groundwater vulnerability mapping." *Canadian Water Resources Journal/Revue Canadienne des Ressources Hydriques* 18(1): 25–37.

Su, Y., L. Zhu et al. (2009). Shallow groundwater quality monitoring and assessment in Northern Ordos cretaceous artisan basin, China. *Bioinformatics and Biomedical Engineering, 2009. ICBBE 2009. 3rd International Conference on.* Beijing, China, IEEE.

Teso, R. R., M. P. Poe et al. (1996). "Use of logistic regression and GIS modeling to predict groundwater vulnerability to pesticides." *Journal of Environmental Quality* 25(3): 425–432.

Tiktak, A., J. J. T. I. Boesten et al. (2006). "Mapping ground water vulnerability to pesticide leaching with a process-based metamodel of EuroPEARL." *Journal of Environmental Quality* 35(4): 1213–1226.

Troiano, J., C. Nordmark et al. (1997). "Profiling areas of ground water contamination by pesticides in California: Phase ii–Evaluation and modification of a statistical model." *Environmental Monitoring and Assessment* **45**(3): 301–319.

Troiano, J., F. Spurlock et al. (1999). *Update of the California Vulnerability Soil Analysis for Movement of Pesticides to Ground Water*. Sacramento, CA, Environmental Monitoring and Pest Management Branch California Department of Pesticide Regulation.

Tseng, C. H., C. K. Chong et al. (2003). "Long-term arsenic exposure and ischemic heart disease in arseniasis-hyperendemic villages in Taiwan." *Toxicology Letters* **137**(1–2): 15–21.

Tziritis, E. and N. Evelpidou (2011). Intrinsic vulnerability assessment using a modified version of the PI Method: A case study in the Boeotia region, Central Greece. *Advances in the Research of Aquatic Environment*. N. Lambrakis, G. Stournaras, and K. Katsanou (Eds.), Berlin, Germany, Springer, pp. 343–350.

Umar, R., I. Ahmed et al. (2009). "Mapping groundwater vulnerable zones using modified DRASTIC approach of an alluvial aquifer in parts of central Ganga plain, Western Uttar Pradesh." *Journal of the Geological Society of India* **73**(2): 193–201.

US Geological Survey (2014). "Contaminants found in groundwater." from https://water.usgs.gov/edu/groundwater-contaminants.html (accessed 25 June 2014).

USEPA (2014). "Ground water and drinking water." from ss://water.epa.gov/drink/index.cfm (accessed 25 June 2014).

USGS (2000). *Arsenic in Ground-Water Resources of the United States*. Reston, VA, U.S. Geological Survey.

Vías, J. M., B. Andreo et al. (2005). "A comparative study of four schemes for groundwater vulnerability mapping in a diffuse flow carbonate aquifer under Mediterranean climatic conditions." *Environmental Geology* **47**(4): 586–595.

Vías, J. M., B. Andreo et al. (2006). "Proposed method for groundwater vulnerability mapping in carbonate (karstic) aquifers: The COP method." *Hydrogeology Journal* **14**(6): 912–925.

Vrba, J. and A. Zaporozec (1994). *Guidebook on Mapping Groundwater Vulnerability (IAH International Contribution for Hydrogegology)*. Rotterdam, the Netherlands, A. A. Balkema Publishers.

Wang, J., W. Sun et al. (2010). Study on the groundwater resources distribution in the old industrial base of Northeast China with GIS. *Bioinformatics and Biomedical Engineering (iCBBE), 2010 4th International Conference on*. Chengdu, China, IEEE.

Wu, J. and B. A. Babcock (1999). "Metamodeling potential nitrate water pollution in the Central United States." *Journal of Environmental Quality* **28**: 13.

Wu, T., M. Yang et al. (2011). Assessment of groundwater specific vulnerability in Guangzhou based on fuzzy comprehensive evaluation. *Water Resource and Environmental Protection (ISWREP), 2011 International Symposium on*. Xi'an, China, IEEE.

Yang, M., H. Xie et al. (2011). Assessment of groundwater intrinsic vulnerability in Guangzhou. *Water Resource and Environmental Protection (ISWREP), 2011 International Symposium on*. Xi'an, China, IEEE.

Zwahlen, F. (2004). *Vulnerability and risk mapping for the protection of carbonate (karst) aquifers: Final report*. Luxembourg, UK, European Commission, Directorate-General for Research and Innovation.

2

Assessment of Effectiveness of DRASTIC Model

Groundwater contamination has been increasing in the Fatehgarh Sahib district in Punjab, India since its establishment in 1992. Fatehgarh Sahib is a major farming region where the number of tube wells used for irrigation has been increasing at an exponential rate. As a result, the depth of groundwater has been declining severely in the last 10 years. The rate of decline is so severe that the latest reported rate of decline is estimated to be 20–90 cm per year (Saigal 2007). According to the report of Ministry of Micro Small and Medium Enterprises, the quality of water in shallow aquifers has been found to be contaminated to the maximum permissible extent of several heavy metal ions and some toxic elements (Government of India, Ministry of MSME 2011). As industrialization has increased in Fatehgarh Sahib, the discharge of industrial effluents has been contributing significantly to groundwater contamination. Though water has been reasonably suitable for the purpose of drinking and domestic activities, there have been several instances near Mandi Gobindgarh (the industrial town of Fatehgarh Sahib) where groundwater has been found to be polluted with heavy metal ions like zinc, lead, copper, etc. due to industrial pollution (Saigal 2007).

Groundwater vulnerability assessment is an empirical method of determining the risk that groundwater in a particular area will become contaminated based on a number of physical parameters that control the movement of contaminants through the vadose zone to the water table. The vulnerability map that is generated from this process can assist in town planning, in the establishment of industries, special economic zones, and delineation of zones in the study area which could be used for several purposes like residential, agricultural, or industrial land uses. Such maps facilitate the government bodies to plan an effective strategy for groundwater monitoring. There have been many models developed for the assessment of groundwater contamination vulnerability and each model is unique in the sense that it is applicable to a particular type of geo-environmental

This chapter is based on **Prashant Kumar et al.**, Assessment of the Effectiveness of DRASTIC in Predicting the Vulnerability of Groundwater to Contamination: A Case Study from Fatehgarh Sahib district in Punjab, India, Environmental Earth Sciences, Springer, Volume 75, May 2016, Issue 10, Pages 1–13 **(Impact Factor- 1.76)**.

conditions (Gogu and Dassargues 2000; Kumar et al. 2015). No vulnerability assessment model is generic enough which can cater to the needs of all kinds of geological environments. Each study area has its own characteristics. The vulnerability models have limitations in the sense that they are qualitative model and doesn't consider the effect of pollution type and the physical attenuation process (Kumar et al. 2015). Also, groundwater flow conditions and the transport properties of subsurface are not captured in vulnerability assessment models.

Of these models, DRASTIC (Aller et al. 1987) is the most widely used model (Thirumalaivasan et al. 2003; Babiker et al. 2005; Pathak et al. 2008; Shirazi et al. 2012; Kaliraj, et al. 2014; Prasad and Shukla 2014; Ghosh et al. 2015; Kardan Moghaddam et al. 2015; Khodabakhshi et al. 2015; Pacheco et al. 2015). It is a theoretical model which tries to discriminate the higher vulnerable zones from lower ones using a number of hydrogeological parameters (Duarte et al. 2015). Even though it is claimed to be the most reliable method in literature, many of the parameters are selected on a subjective basis (Shirazi et al. 2012; Kumar et al. 2015). At the same time, the accuracy of results obtained from groundwater vulnerability models are subjected to many challenges. In this chapter, it is observed that the predictions of groundwater vulnerability made by DRASTIC model may not accurately represent actual groundwater quality measured in tube wells. There are a number of instances where groundwater monitoring confirmed the presence of hazardous elements at some locations that were not considered to have a high risk of contamination by the DRASTIC model. After all, such studies can never replace on-site inspection. They are totally empirical in nature (Panagopoulos et al. 2006; Jang et al. 2015). The chapter examines the accuracy of the DRASTIC vulnerability model based on the results of groundwater monitoring and recommends the inclusion of land use and additional hydrogeological parameters other than already considered under DRASTIC model are also taken up into consideration (Javadi et al. 2010; Kazakis and Voudouris 2015; Neshat et al. 2015).

Methodology

DRASTIC model is one of the most widely used model across the world. Though it is not a generic model, it still gives a good estimate of vulnerability index irrespective of the study area. However, as such studies are location-specific studies, there is always a scope for better model incorporating several other hydrogeological parameters and anthropogenic factors which are not currently considered in the DRASTIC model. This study is with respect to DRASTIC model application to the study area and its accuracy assessment.

GIS for Hydrological Investigations

The large scale agricultural development, irrigation, urbanization, and industrialization have direct impact of the groundwater reservoirs such as contamination of groundwater and depletion of water table and has become a very serious issue for any growing country and region. Identification and mapping of potable groundwater sources which are sustainable for longer periods is a critical issue in the supply of drinking and irrigation water to the rural habitations across the world. Remote sensing and geographical information system (GIS) play a significant role in making it possible to undertake studies on sustainable water use, watershed development planning, including that of sustainable groundwater use, and recharge. Ground prospects zonation means identifying and mapping the prospective groundwater zones in an area by qualitative assessment of the controlling and indicative parameters. Qualitative interpretation of features of groundwater interest is largely possible through remote sensing and ground checks. Remotely sensed data are used for lithological and structural mapping of water-bearing formations, geomorphological mapping, drainage basin analysis, determining flood plain areas and areas of lakes/swamps, determining locations in the field, and other groundwater indicators.

As far as groundwater quality zonation is concerned, GIS is being increasingly used to: (1) prepare the databases about the quality parameters (measured using conventional methods), (2) prepare and analyze the quality parameter maps for spatial representation of groundwater quality and for providing information about the parameters causing the problem or pollution, and (3) map the zones vulnerable to groundwater pollution due to various reasons, viz., industrial, agricultural, municipal, etc. For the integrated analysis of different parameters, generally weighted index model is used which is easily implementable in GIS domain.

Study Area

Fatehgarh Sahib district is the pilot study area. It is one of the smallest districts of Punjab state and is located in the south-eastern part of Punjab. It lies between 30° 25' 00" to 30° 45' 45" north latitude and 76° 04' 30" to 76° 35' 00" east longitude. The total geographical area of the district is 1147 km². The study area is divided into five blocks/divisions, viz., Khamanon, Bassi Pathana, Khera, Sirhind, and Amloh. The georeferenced-mosaic toposheet of the study area is shown in Figure 2.1.

The major canals passing through Fatehgarh Sahib are Bhakra canal, Sirhind canal, and Satluj Yamuna Link canal as shown in Figure 2.2. Bhakra canal passes through the middle of the district in a north to south direction. There are four major distributaries—1L and 2L distributaries flowing towards eastern and south-eastern part, respectively and 1R and 2R distributaries flowing towards western and south-western part, respectively, of the district,

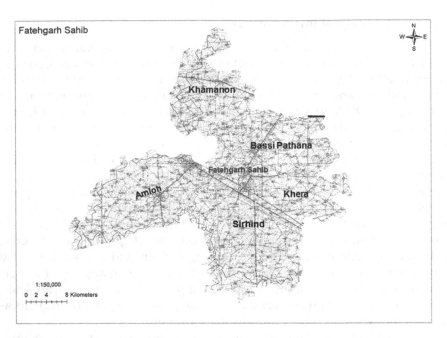

FIGURE 2.1
Geographical boundary of Fatehgarh Sahib district of Punjab.

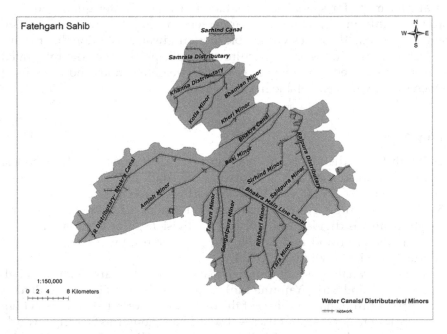

FIGURE 2.2
Network of water canals/distributaries/minors in Fatehgarh Sahib district.

bifurcate from Bhakra canal in the central part of the district. These distributaries further develop several minor channels such as Amloh channel in the south-western part of the district, the Ritkheri & Tara minors in the south, and the Saidpura channel in the east. There are three major distributaries (Samrala, Khanna, and Gobindgarh) in the northern part of the district which originate from Sirhind Canal. Among the minor channels in northern part are Bhamian, Kheri, Kotla, Latheri, and Sangtpura. Sirhind canal passes through the western part of the district, and it flows from north-west to south-east direction.

Fatehgarh Sahib is an agrarian district with some industrial activity associated with steel plants in Mandi Gobindgarh. Most of the district lies in a fertile alluvial plain with many dug wells and tube wells for an extensive irrigation system. The irrigation is also provided by distributaries and minor channels of the Bhakra canal. There is excessive groundwater use in the district because of deepening of bore wells by farmers and replacement of centrifugal pumps by submersible one (Saigal 2007). As a result of overexploitation, the quality and quantity of groundwater has been progressively declining.

The five blocks in Fatehgarh Sahib have been further divided into villages which constitute the complete study area—Fatehgarh Sahib district.

DRASTIC Model

DRASTIC model is discussed in detail under Paramatic Models in Section "**Pragmatic**" in Chapter 1. DRASTIC model has been implemented in the study area—Fatehgarh Sahib district of Punjab in conjunction with geographic information system to study several meteorological conditions.

Estimation of Hydrogeological Parameters

The seven parameters of DRASTIC model have been derived in the following manner.

Depth of Water Table

The depth of the water table is similar in both the pre- and post-monsoon seasons in the district. This is evident from Figures 2.3 and 2.4 which indicate that over the period 2008–2013, the average depth of water table varied from 14.81 m below ground level (mbgl) to 29.5 m below ground level in the pre-monsoon season (June) and from 14.45 m below ground level (mbgl) to 30.1 m below ground level in the post-monsoon season (October). The pre-monsoon average depth of water table (in mbgl) in five blocks of Fatehgarh Sahib are 20.88 (Amloh), 19.11 (Bassi Pathana), 17.05 (Khamanon), 18.92 (Khera), and 18.26 (Sirhind). The post-monsoon average depth of water table (in mbgl) in five blocks of Fatehgarh Sahib are 20.79 (Amloh), 18.94 (Bassi Pathana),

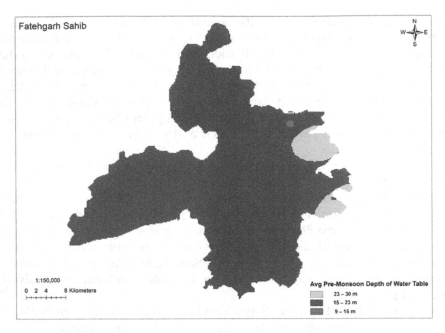

FIGURE 2.3
Profile of average pre-monsoon depth of water table in Fatehgarh Sahib.

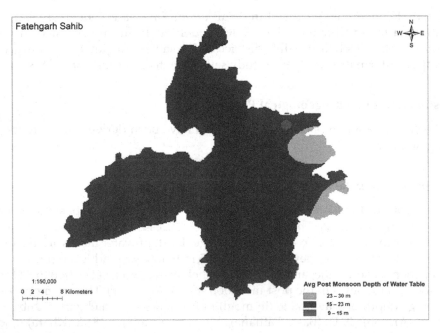

FIGURE 2.4
Profile of average post-monsoon depth of water table in Fatehgarh Sahib.

16.80 (Khamanon), 17.97 (Khera), and 18.10 (Sirhind). It is to be noticed from Figure 2.3 that eastern part of the district covering few villages under blocks Bassi Pathana and Khera has the maximum depth of water table ranging from 22.86 to 30.48 mbgl in the pre-monsoon season. The situation remain almost same except few villages in the post-monsoon season also as seen from Figure 2.4. Most of villages under five blocks have similar depth of water table in the range from 15.24 to 22.86 mbgl.

Values corresponding to the depth of water table have been divided into three classes and respective ratings have been assigned in the final calculation of vulnerability index values to assess the pollution potential.

It is expected that post-monsoon depth of water table should be less from the earth surface in absolute depths which is not the case here as evident from Figure 2.4 due to excessive evaporation and surface runoff. Also, the large rate of groundwater abstraction of water due to the excessive number of tube wells for irrigation plays a vital role in countering the recharge from the rainfall during the monsoon season which limits the extent of the recovery of the water table.

Net Recharge

Recharge of groundwater can take place from many sources such as rainfall, canal, river, irrigation, tanks, ponds, and water conservation structures. Net recharge in Fatehgarh Sahib district was calculated using the governing formula as given in Equation (2.1).

$$\text{Net recharge} = R_{\text{Rainfall}} + R_{\text{Canal}} + R_{\text{Surface water irrigation}}$$
$$+ R_{\text{Tanks and ponds}} + R_{\text{Groundwater irrigation}} + R_{\text{Water conservation structure}}, \tag{2.1}$$

where, R_{Rainfall} = recharge due to rainfall, R_{Canal} = recharge due to seepage from canals, $R_{\text{Surface water irrigation}}$ = recharge from irrigation done using surface water, $R_{\text{Tanks and ponds}}$ = recharge from ponds and tanks, $R_{\text{Groundwater irrigation}}$ = recharge from irrigation done using groundwater, and $R_{\text{Water conservation structure}}$ = recharge from various water storage structures.

The magnitude of recharge from rainfall and other sources during the monsoon and non-monsoon seasons for years 2008–2009 and 2010–2011 are shown in Table 2.1. These values have been taken from the reports of DYNAMIC GROUNDWATER RESOURCES OF PUNJAB STATE 2009 and 2011 (Water Resources and Environment Directorate Punjab 2009, 2011).

It is to be observed that there is not much variation in total annual groundwater recharge in both the years. As the recharge is more than 250 mm in all the five blocks of Fatehgarh Sahib district, therefore a homogeneous rating of 9 has been assigned to net recharge parameter in final vulnerability index assessment (Aller et al. 1987; Kumar et al. 2015).

TABLE 2.1

Groundwater Recharge in Fatehgarh Sahib in Years 2008–2009 and 2010–2011

Year	Blocks	During Monsoon Season (in ham) Recharge from Rainfall and Other Sources	During Non-Monsoon Season (in ham) Recharge from Rainfall and Other Sources	Total Annual Groundwater Recharge (in ham)	Natural Discharges (in ham)	Net Annual Groundwater Availability (in ham)	Recharge in meters = (Net Groundwater Availability in ham/area in ha)
2008–2009	Khera	7621	2384	10005	1001	9005	0.4980641
	Sirhind	13757	3714	17471	1747	15724	0.4222341
	Amloh	9915	3359	13274	1327	11946	0.5381081
	Bassi Pathana	7719	2324	10043	1004	9039	0.4846648
	Khamanon	7151	1506	8657	866	7791	0.5026451
	Total	46163	13287	59450	5945	53505	0.4791349
2010–2011	Khera	7627	2369	9996	1000	8996	0.4975663
	Sirhind	13799	3698	17497	1750	15747	0.4228517
	Amloh	9995	3351	13346	1335	12011	0.5410360
	Bassi Pathana	7914	2312	10226	1023	9203	0.4934584
	Khamanon	7349	1498	8847	885	7962	0.5136774
	Total	46684	13228	59912	5993	53919	0.4828423

Aquifer Type

The physical characteristics of an aquifer together with its hydraulic conductivity also have a large influence on the level of vulnerability determined by the DRASTIC model. Fatehgarh Sahib district is underlain by an extensive shallow aquifer in alluvial sediments (Saigal 2007; Water Resources and Environment Directorate Punjab 2009, 2011). According to Park et al. (2007) alluvial deposits are very good for supplying water to many rural agricultural areas due to the water holding characteristics of these materials. Both the hydraulic conductivity and the porosity of sediments materials can affect in a shallow alluvial aquifer and can have a large influence on its vulnerability to contamination. The bigger the openings/fracture size, the higher will be the pollution potential as the particular aquifer will have a higher permeability and lower attenuation potential (Umar et al. 2009). In this study, a rating of 8 has been assigned for the aquifer type in the final calculation of vulnerability index of contamination because area aquifer material is almost homogeneous throughout the district. The alluvial aquifer in the Fatehgarh Sahib district has a thickness of about 20 m and is comprised of clay and fine sand.

Soil Type

Soil comprises the upper 1–2 m part of the weathered profile that is most affected by biological processes. The structure and composition of a soil profile has a large influence on the transport of a contaminant from the land surface through the vadose zone to groundwater. The DRASTIC vulnerability index depends on the type of soil present in the study area and on the average grain size of mineral grains in the soil profile. Soils in the Fatehgarh Sahib district are mainly comprised of loams, therefore an average uniform rating of 5 has been assigned to the soil type parameter in the computation of final vulnerability index for contamination (Aller et al. 1987; Kumar et al. 2015).

Topography

The elevation and slope of the land surface in the study area is shown in the Figure 2.5 and indicates that Fatehgarh Sahib is largely a flat region. Areas with higher slope have a larger contribution to surface runoff because they shed water from rainfall more rapidly than flat areas and therefore have a lower capacity to allow water infiltration than flat area. As a result, the risk of contaminant infiltration is lower in areas with a steep slope than areas which are largely flat. On the other hand, areas with smaller slope contain the water for longer duration of time thereby letting the contaminants pass through the vadose zone to reach the groundwater, subsequently increasing the contaminant infiltration. The topography map has been generated using digital elevation model of Fatehgarh Sahib in GIS software (Aller et al. 1987). From Figure 2.5, it can be seen that major portions of the study area are flat having

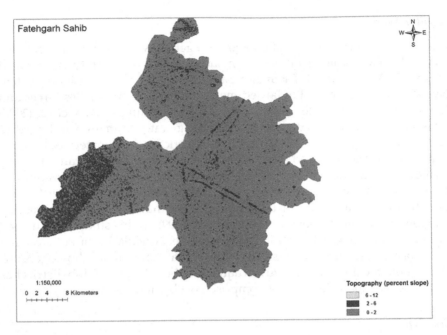

FIGURE 2.5
Topography profile of Fatehgarh Sahib.

slope in the range 0%–2%. Areas located in the western side (Amloh block) of the study area are having slope in the range from 2% to 6%. There are very few areas which are having slope in the range of 6–12 as shown in cyan color. The slope values have been divided into three classes and corresponding rating values have been assigned.

Lithological Structures and Vadose Zone Assessment

The portion of the weathered profile or fractured rock between the land surface and the water table is known as the vadose zone. The extent to which contaminants reach the water table depends on the degree to which they are attenuated by biological and chemical processes in the vadose zone. In general, the extent of contaminant attenuation increases with increasing thickness of the vadose zone. Data from 24 boreholes (Figure 2.6) have been used to determine the thickness of the vadose zone in the study area.

For each of 24 drilling locations, the impact of vadose zone has been assessed from earth's surface to the depth of the water table. Lithology till the occurrence of water table has been considered. Also, the ratings for the effect of the vadose zone as indicated by Aller et al. (1987) were not used in the vulnerability model used, another the weighted average value of ratings has been taken for better impact assessment.

FIGURE 2.6
Sites/villages in Fatehgarh Sahib district for drilling operation.

For example, consider the case of Chandiala village under Khamanon block of Fatehgarh Sahib district, the drill data corresponding to this village are as given Table 2.2.

The empirical formula to calculate the impact of vadose zone is give below:

$$\text{Impact of vadose zone} = \left(\frac{1}{\text{total depth}} \right) \sum_{i=1}^{n} \text{depth}_{L_i} \times \text{rating}_{L_i}, \quad (2.2)$$

where, depth_{L_i} = depth of the particular lithological unit such as sand, clay, stone, etc., rating_{L_i} = rating given to the particular lithological unit by Aller et al. (1987), total depth = depth of the water table from earth's surface.

From the Figure 2.7, it can be seen that the impact of vadose zone is higher in the western part of the study area (Amloh Block) and in some areas corresponding to the central regions of Fatehgarh Sahib. The eastern part of the study area has the lowest impact of the vadose zone (i.e., the depth of the water table is largest in this area). The rest of the study area has moderate impact of the vadose zone. The impact of the vadose zone contributes immensely in the final calculation of the vulnerability index. The higher the impact, the higher will be the pollution potential as the governing equation for the vulnerability assessment is linear and additive in nature.

TABLE 2.2

Lithological Structure in Chandiala Village of
Khamanon Block in Fatehgarh Sahib

Fatehgarh Sahib		Chandiala, Khamanon
Depth (m)	Depth (m)	Lithology
0.00	0.50	Surface clay
0.50	1.00	Surface clay
1.00	1.50	Clay
1.50	2.00	Clay
2.00	3.00	Clay
3.00	4.00	Clay
4.00	5.00	Clay
5.00	6.00	Clay
6.00	7.00	Fine sand
7.00	8.00	Fine sand
8.00	9.00	Fine sand
9.00	10.00	Fine sand
10.00	11.00	Fine sand
11.00	12.00	Fine sand
12.00	13.00	Fine sand
13.00	14.00	Fine sand
14.00	15.00	Fine sand
15.00	16.00	Fine to medium sand
16.00	17.00	Fine to medium sand

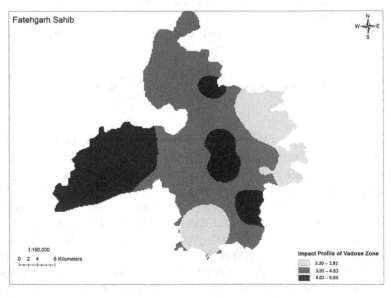

FIGURE 2.7
Impact profile of vadose zone in Fatehgarh Sahib.

Hydraulic Conductivity

The saturated hydraulic conductivity of aquifer sediments is a measure of the volume of water that can flow through a unit cross sectional area of the aquifer per unit time under a specified hydraulic gradient. It also regulates the rate at which soluble contaminants are moved in groundwater flow. The magnitude of the hydraulic conductivity of aquifer sediments increases as the amount and degree of interconnection of void spaces within the aquifer also increases (Umar et al. 2009). As Fatehgarh Sahib district is a single physiographic unit and the whole region falls in alluvial plain, therefore, hydraulic conductivity values do not show a large degree of variation. Also, clays are the major dominating part of the subsurface; therefore, a homogeneous rating of 1 has been assigned to hydraulic conductivity parameter in final estimation of groundwater vulnerability index.

Implementation of DRASTIC Model and Analysis

Vulnerability Map

Vulnerability indices have been calculated using the empirical formula given in Equation (1.1). The vulnerability map generated out of these vulnerability indices is shown in Figure 2.8. The vulnerability index values range

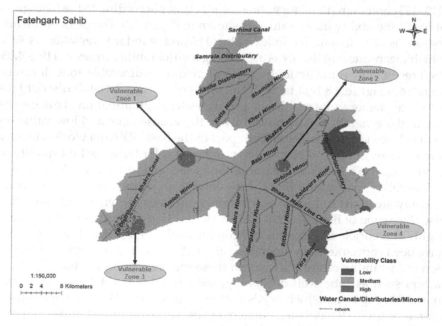

FIGURE 2.8
Groundwater vulnerability map of Fatehgarh Sahib.

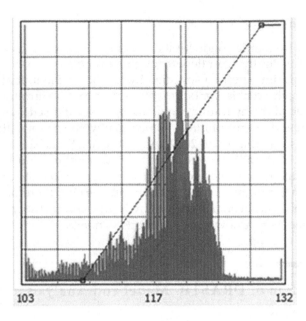

FIGURE 2.9
Histogram of vulnerability index values of Fatehgarh Sahib.

from 103 to 132. These values have been classified into low (100–110), medium (110–125), and high (125–140) vulnerability classes according to the histogram of the vulnerability index values as shown in Figure 2.9. The peak of the histogram is at vulnerability index value 119 and standard deviation is 4.03, which means most of the areas are having a vulnerability index of 119 ± 4.03. As a result, they come under the class of medium vulnerable zone. It can be seen from Figure 2.8 that most of the areas of Fatehgarh Sahib district falls under moderate vulnerability of groundwater contamination. This can be due to the generally low permeability of the vadose zone and low values of hydraulic conductivity. The eastern part of the Bassi Pathana block and very few portions of Sirhind block shown in violet color have the best quality of groundwater. There is no contamination as such in those places. A perusal of vulnerability map shows that the western, central, and south-eastern parts of study area falls in higher vulnerability zones as shown in red color. It is also the region of Bhakra canal and its 1R and 1L distributaries. This zone is highly susceptible to groundwater contamination. Higher vulnerable zones have been earmarked as vulnerable zone 1, 2, 3, and 4. This high contamination vulnerability of groundwater in these regions can be attributed to local factors such as tube well deepening, industrial units establishments, and many other reasons which needs further exploration. There are no natural source of toxic elements (elaborated in the validation part), thus, its presence in the groundwater hints the leaching of nitrogenous organic fertilizers, sludge disposal, human/animal extra disposals, etc. The high frequency of irrigation can also be a governing factor in contamination of groundwater with surface pollutants. Groundwater pollution can also be caused by plunging of the industrial wastes into the open ground, which results into stagnation; leading to various soil and groundwater issues.

Validation

The result of the study shows that most of the regions of Fatehgarh Sahib falls under moderate vulnerability except for few zones of higher vulnerability. In order to validate the results, the tube well sample data from all the four vulnerable zones 1, 2, 3, and 4 (from Department of Water Supply & Sanitation Department, Mohali, Phase–6, Punjab, India and its laboratory analysis reports of year 2013) have been evaluated. The location of the samples taken for chemical analysis is shown in Figure 2.10, and it confirms the presence of highly toxic elements such as selenium, lead, fluoride, nitrate, and iron in these four vulnerable zones which accounts for a reasonably good estimation of groundwater vulnerability in Fatehgarh Sahib district.

The acceptable limits of selenium, lead, nitrate, fluoride, and iron are 0.01, 0.01, 45, 1.0, and 0.3 mg/L, respectively. Vulnerable zone 1 is dominated by presence of excessive iron. Vulnerable zone 2 is full of fluoride, selenium, and excessive iron. Vulnerable zone 3 has instances of lead and fluoride, and vulnerable zone 4 is contaminated with excessive iron and fluoride.

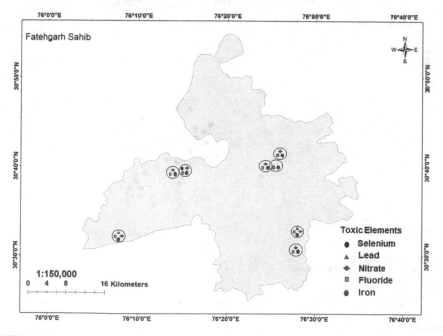

FIGURE 2.10
Toxic elements at selected sites for validation.

It is observed that though DRASTIC model has been useful in localizing the some of the higher vulnerable zones, but at the same time its accuracy can always be debated as far as a true/exact picture of groundwater vulnerability is concerned. The main reason for such deviation is that it is a theoretical model which depends heavily on different hydro-geological parameters. DRASTIC uses seven parameters which is good enough for a rough approximate estimation of mapping of contaminated zones. In order to enhance the accuracy, the model will have to be modified accordingly (Neshat et al. 2014), one will have to incorporate parameters other than those already considered under DRASTIC. Anthropogenic activities such as land use parameters, urban settlements, mining and exploration activities, etc. have significant effects on the depth of water and groundwater contamination (Sener and Davraz 2013; Pórcel et al. 2014; Lavoie et al. 2015; Şener and Şener 2015).

In the study, it is observed the tube well sample data from other places also; other than those considered in validation of the vulnerability map, wherein reasonably enough amount of toxic elements such as lead, selenium, nitrate, fluoride, and iron have been found, but the model under study has not classified those areas as severely vulnerable. Figure 2.11 shows the presence of fluoride beyond the acceptable limit in several villages of Fatehgarh Sahib district.

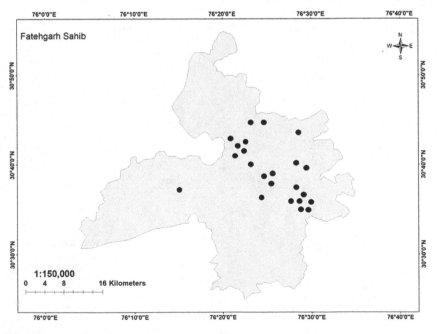

FIGURE 2.11
Fluoride contaminated villages of Fatehgarh Sahib.

Conclusion and Further Recommendations

Fatehgarh Sahib district is a moderately contaminated zone according to the application of DRASTIC model. Excessive use of tube wells has caused groundwater contamination, and the application of polluted groundwater for cultivation and farming activities has also increased the risk that toxic elements like selenium could enter local food webs.

It has already stated that there is no generic groundwater vulnerability assessment model which can account for all potential sources and pathways for groundwater contamination (Kumar et al. 2015). In this study, some deviations in ground realities and theoretical vulnerability map generated using DRASTIC model have been observed. Nevertheless, DRASTIC does give a rough estimate of vulnerable zones. Therefore, it is concluded that validation, accuracy assessment, and precise mapping of such studies are always a challenging task. In order to get most accurate or precise mapping, one must include as many environmental parameters as one can. The interaction of these parameters is altogether a different research and the outcomes of such interactions significantly affect the environmental occurrences such as groundwater.

References

Aller, L., T. Bannet et al. (1987). *DRASTIC: A Standardized System to Evaluate Ground Water Pollution Potential Using Hydrogeologic Settings*. Worthington, OH, National Water Well Association.

Babiker, I. S., M. A. Mohamed et al. (2005). "A GIS-based DRASTIC model for assessing aquifer vulnerability in Kakamigahara Heights, Gifu Prefecture, central Japan." *Science of the Total Environment* **345**(1–3): 127–140.

Duarte, L., A. C. Teodoro et al. (2015). "A dynamic map application for the assessment of groundwater vulnerability to pollution." *Environmental Earth Sciences* **74**(3): 2315–2327.

Ghosh, A., A. Tiwari et al. (2015). "A GIS based DRASTIC model for assessing groundwater vulnerability of Katri Watershed, Dhanbad, India." *Modeling Earth Systems and Environment* **1**(3): 1–14.

Gogu, R. C. and A. Dassargues (2000). "Current trends and future challenges in groundwater vulnerability assessment using overlay and index methods." *Environmental Geology* **39**(6): 549–559.

Government of India, Ministry of MSME (2011). *Brief Industrial Profile of District Fatehgarh Sahib*. Ludhiana, India

Jang, C. S., Lin, C. W., Liang, C. P., & Chen, J. S. (2016). Developing a reliable model for aquifer vulnerability. *Stochastic environmental research and risk assessment*, **30**(1), 175-187.

Javadi, S., N. Kavehkar et al. (2010). "Modification of DRASTIC model to map ground-water vulnerability to pollution using nitrate measurements in agricultural areas." *Journal of Agricultural Science and Technology* **13**: 239–249.

Kaliraj, S., N. Chandrasekar et al. (2014). "Mapping of coastal aquifer vulnerable zone in the south west coast of Kanyakumari, South India, using GIS-based DRASTIC model." *Environmental Monitoring and Assessment* **187**(1): 1–27.

Kardan Moghaddam, H., F. Jafari et al. (2015). "Evaluation vulnerability of coastal aquifer via GALDIT model and comparison with DRASTIC index using quality parameters (accepted for publication)." *Hydrological Sciences Journal* **62**: 137–146.

Kazakis, N. and K. S. Voudouris (2015). "Groundwater vulnerability and pollution risk assessment of porous aquifers to nitrate: Modifying the DRASTIC method using quantitative parameters." *Journal of Hydrology* **525**: 13–25.

Khodabakhshi, N., G. Asadollahfardi et al. (2015). "Application of a GIS-based DRASTIC model and groundwater quality index method for evaluation of groundwater vulnerability: A case study, Sefid-Dasht." *Water Science and Technology: Water* **15**(4): 784–792.

Kumar, P., B. K. Bansod et al. (2015). "Index-based groundwater vulnerability mapping models using hydrogeological settings: A critical evaluation." *Environmental Impact Assessment Review* **51**: 38–49.

Lavoie, R., F. Joerin et al. (2015). "Integrating groundwater into land planning: A risk assessment methodology." *Journal of Environmental Management* **154**: 358–371.

Neshat, A., B. Pradhan et al. (2014). "Estimating groundwater vulnerability to pollution using a modified DRASTIC model in the Kerman agricultural area, Iran." *Environmental Earth Sciences* **71**(7): 3119–3131.

Neshat, A., B. Pradhan et al. (2015). "Risk assessment of groundwater pollution using Monte Carlo approach in an agricultural region: An example from Kerman Plain, Iran." *Computers, Environment and Urban Systems* **50**: 66–73.

Pacheco, F. A. L., L. M. G. R. Pires et al. (2015). "Factor weighting in DRASTIC modeling." *Science of the Total Environment* **505**: 474–486.

Panagopoulos, G., A. Antonakos et al. (2006). "Optimization of the DRASTIC method for groundwater vulnerability assessment via the use of simple statistical methods and GIS." *Hydrogeology Journal* **14**(6): 894–911.

Park, Y. H., S. J. Doh et al. (2007). "Geoelectric resistivity sounding of riverside alluvial aquifer in an agricultural area at Buyeo, Geum River watershed, Korea: An application to groundwater contamination study." *Environmental Geology* **53**(4): 849–859.

Pathak, D. R., A. Hiratsuka et al. (2008). GIS based fuzzy optimization method to groundwater vulnerability evaluation. *Bioinformatics and Biomedical Engineering, 2008. ICBBE 2008. The 2nd International Conference on.* Shanghai, China, IEEE.

Pórcel, R. A. D., C. Schüth et al. (2014). "Land-use impact and nitrate analysis to validate DRASTIC vulnerability maps using a GIS platform of pablillo river basin, Linares, NL, Mexico." *International Journal of Geosciences* **5**(12): 1468.

Prasad, K. and J. Shukla (2014). "Assessment of groundwater vulnerability using GIS-based DRASTIC technology for the basaltic aquifer of Burhner watershed, Mohgaon block, Mandla (India)." *Current Science* **107**(10): 1649.

Saigal, S. K. (2007). *Ground Water Information Booklet Fatehgarh Sahib District, Punjab.* Chandigarh, India, Central Groundwater Board North Western Region.

Sener, E. and A. Davraz (2013). "Assessment of groundwater vulnerability based on a modified DRASTIC model, GIS and an analytic hierarchy process (AHP) method: The case of Egirdir Lake basin (Isparta, Turkey)." *Hydrogeology Journal* **21**(3): 701–714.

Şener, E. and Ş. Şener (2015). "Evaluation of groundwater vulnerability to pollution using fuzzy analytic hierarchy process method." *Environmental Earth Sciences* **73**(12): 8405–8424.

Shirazi, S. M., H. M. Imran et al. (2012). "GIS-based DRASTIC method for groundwater vulnerability assessment: A review." *Journal of Risk Research* **15**(8): 991–1011.

Thirumalaivasan, D., M. Karmegam et al. (2003). "AHP-DRASTIC: Software for specific aquifer vulnerability assessment using DRASTIC model and GIS." *Environmental Modelling & Software* **18**(7): 645–656.

Umar, R., I. Ahmed et al. (2009). "Mapping groundwater vulnerable zones using modified DRASTIC approach of an alluvial aquifer in parts of central Ganga plain, Western Uttar Pradesh." *Journal of the Geological Society of India* **73**(2): 193–201.

Water Resources and Environment Directorate Punjab. (2009). *Dynamic Ground Water Resources of Punjab State*. Chandigarh, India, Central Groundwater Board.

Water Resources and Environment Directorate Punjab. (2011). *Dynamic Ground Water Resources of Punjab State*. Chandigarh, India, Central Groundwater Board.

3

Multi-criteria Evaluation of Hydrogeological and Anthropogenic Parameters

One of the limitations of the fundamental DRASTIC model is that it doesn't consider several other important geo-environmental parameters which are very significant from the perspective of accuracy of groundwater vulnerability map. One of the important such parameter is anthropogenic factors. Urban settlements play a significant role in groundwater contamination (Sener and Davraz 2013; Martín del Campo et al. 2014; Ouedraogo et al. 2016). Due to the fixed ratings and weights of DRASTIC parameters, the accuracy of groundwater vulnerability map generated is always challenged depending upon the local environmental conditions (Kumar et al. 2016). Further, the fundamental DRASTIC model doesn't explain any scientific reason for the assignment of the particular ratings and weights. These ratings and weights are based on the experience and opinions of the experts. This brings human subjectivity in assignment of the ratings and weights. To alleviate such issues, there have been many modifications of DRASTIC model in terms of hydrogeological parameters selection, reduction, and addition (Chenini et al. 2015; Kazakis and Voudouris 2015; Sadat-Noori and Ebrahimi 2015). The scientific basis of assigning the ratings and weights has also been explored by adopting various optimization techniques AHP (analytic hierarchy process)-DRASTIC (Sener and Davraz 2013; Kang et al. 2016; Sahoo et al. 2016a,b; Shen et al. 2016), Grey AHP-DRASTIC (Sahoo et al. 2016a,b), Fuzzy-DRASTIC (Shouyu and Guangtao 2003; Rezaei et al. 2013; Iqbal et al. 2015; Arezoomand Omidi Langrudi et al. 2016), and so on. In most of the cases of AHP-DRASTIC implementation, weights have been kept as per what Aller et al. (1987) have suggested and only the ratings have been optimized according to the regional characteristics of study area. Further, one of the parameters of DRASTIC—vadose zone has not been considered in detail in most of the earlier studies.

This chapter deals with the optimization of ratings and weights of DRASTIC parameters and the relative comparison of groundwater vulnerability maps generated by AHP-DRASTICL, Modified DRASTIC, and the fundamental DRASTIC and associated accuracies have been assessed. The novelty of the work lies in the fact that a scientific consideration to the anthropogenic factors along with the DRASTICL parameters has been given

*This chapter is based on **Prashant Kumar et al.**, Multi-Criteria Evaluation of Hydro-geological and Anthropogenic Parameters for the Groundwater Vulnerability Assessment, Environmental Monitoring and Assessment, Springer, Volume 189, Nov 2017, Issue 11, Pages 1–24 (**Impact Factor- 1.687**).*

and further, optimization of both the ratings and weights has been done by using a very comprehensive multi-criteria evaluation technique (AHP) for better accuracy assessment by building a hierarchy of relative comparisons of hydrogeological parameters and adoptions of various criteria for optimal value assignments. Further, such works are unique in the sense that they are location specific works which are highly influenced by the local environmental conditions. One of the most important aspect of this study is that the impact of vadose zone has been assessed, a very important component for contaminants infiltration from the earth's surface to the groundwater, in a very in-depth manner by giving due consideration to every subsurface layer from the earth surface to the occurrence of groundwater. This results in more realistic assessment of vulnerability of groundwater contamination. The application of AHP for environmental related issues has got a tremendous acceptance across the world as found in the literature (Huang et al. 2011; Mardani et al. 2015). The study area is Fatehgarh Sahib district from Punjab, India which is already declared as a region which is facing groundwater contamination and overexploitation (Saigal 2007; Kumar et al. 2016).

Materials and Methods

The schematic of the methodology is shown below in Figure 3.1.

The research process consists of application of DRASTIC and its derived versions, viz., Modified DRASTIC and AHP-DRASTICL (described later in

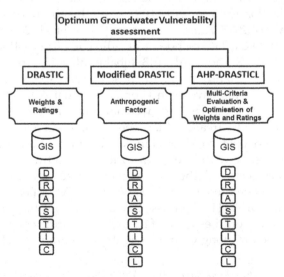

FIGURE 3.1
Methodological steps of the research process.

this section) over a pilot study area (Fatehgarh Sahib, Punjab) to prepare a reasonably accurate groundwater vulnerability map. The study follows the comparative assessment of the groundwater vulnerability maps generated by fundamental DRASTIC model and its derivative models. The modifications done to the derived versions of DRASTIC pertain to the inclusion of anthropogenic factors and the optimization of weights and ratings vectors for a comprehensive assessment of vulnerability of groundwater to contamination.

The accuracy of the different classes of vulnerability shown by Modified DRASTIC and AHP-DRASTICL has been assessed by ten water quality parameters. The exercise of collection of water samples from across the entire study region has been done by Water Supply and Sanitation Department, Mohali, Punjab, India. Water samples from the entire study region were taken in the years 2013 (May, June, July, August) and 2014 (January) from approximately 241 sample points spread uniformly across the study region. Of the water quality parameters, pH and total dissolved solids (TDS) have been measured by electrometric method; alkalinity, hardness, magnesium, sulphate, and calcium have been measured by chemical methods. Iron content has been assessed by inductively coupled plasma optical emission spectrometry (ICP-OES) method as a part of heavy metal ions. Fluoride has been assessed using ION-selective electrochemical method, and nitrate has been measured by UV spectrophotometric method.

Hydrogeological Settings

Alluvial deposits have been found to lie beneath the study area. Subsurface geological constituents range from clay to fine to medium sand to silt to sandstones. As per Groundwater Exploration Committee-97 report, Central Groundwater Board has carried out the groundwater exploration activity in Rasulpur village in Khera Block. No bedrock has been found up to 550 m, thereby making the thickness of the alluvial deposits to be around 550 m (Saigal 2007). Subsurface geological formations found below the earth surface are clay mixed with sand up to 15 m, granular zones up to 30 m, and clay bed up to 20 m, in the increasing order of the depth from the surface. Another clay bed of thickness 30 m is found after a depth of 90–120 m. The thickness of the finer deposits increases after 100 m in the eastern part of the district (Saigal 2007; Kumar et al. 2017). Figure 3.2 shows the hydrogeological settings of the study area and the vertical cross section of the eastern part of study area showcasing the various geological subsurface formations.

The direction of the groundwater flow is from the north-east to the south-west, and they are in the form of flowing ephemeral streams (Singh 1998) as shown in Figure 3.2. Water level elevation varies from 246 meter above mean sea level (amsl) to 266 meter above mean sea level (Saigal 2007). The hydraulic gradient varies from steeper (0.001 in north-east) to gentle (0.00036 in south-west) in the district.

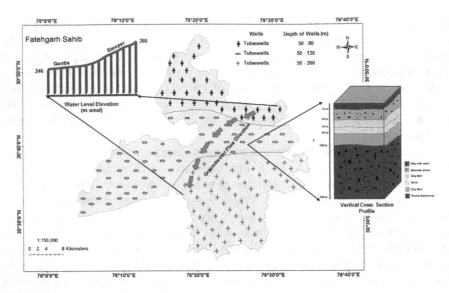

FIGURE 3.2
Hydrogeological settings of the study area.

Data Used

Several hydrogeological data have been used in this study such as depth of water table, recharge, aquifer type, soil type, topography, vadose zone, hydraulic conductivity, and anthropogenic factors. Table 3.1 shows the type of data and its source.

Groundwater Vulnerability Assessment

This section discusses the implementation of DRASTIC, consideration of anthropogenic factors, and multi-criteria evaluation technique for better assessment of vulnerability of groundwater to contamination using DRASTIC.

TABLE 3.1

Hydrogeological Data and Its Source

S. No.	Type of Data	Sources
1	Depth of water table	Department of Irrigation, Punjab, India
2	Net recharge	Central Groundwater Board, Chandigarh, India
3	Aquifer type	Geological Survey of India, Chandigarh, India
4	Soil map	ISRO-IIRS, Dehradun, India
5	Lithological details	Department of Irrigation, Punjab, India
6	Topography	ISRO-IIRS, Dehradun, India
7	Hydraulic conductivity	Central Groundwater Board, Chandigarh, India
8	LULC map	ISRO-IIRS, Dehradun, India
9	Water quality parameters	Department of Water Supply & Sanitation, Punjab, India

DRASTIC Model

The implementation of DRASTIC model and associated results are already discussed in Chapter 2. It is tried to assess the effectiveness of DRASTIC model in the study area Fatehgarh Sahib district and concluded that fundamental DRASTIC model failed to earmark several severe contaminated zones by showing those areas as moderately contaminated zones.

Modified DRASTIC Model (DRASTICL) with Anthropogenic Factors

The fundamental DRASTIC model has its own limitations in the form of accuracy of expressed severity of vulnerability as highlighted by us (Kumar et al. 2016) in the assessment of effectiveness of model. One of the most important parameters which DRASTIC has ignored is the effects of anthropogenic factors (Panagopoulos et al. 2006). Human beings have been using land since ages to meet their requirements. Human interference in the nature has caused an extreme landscape changes (Lambin et al. 2001; Zhang et al. 2015; Ayele et al. 2016). Urban development practices across the world regulate the land dynamics on larger scale (Nourqolipour et al. 2016). Fatehgarh Sahib district has seen significant changes in land use and land cover in the last two decades due to rapid urbanization, developmental work, demography, and agricultural expansion. As per the census of 2011, there has been a change of almost 11.5% in the total population of Fatehgarh Sahib district since the census of 2001, thereby increasing the population density from 456 persons per sq. km. to 523 persons per sq. km (Directorate of Census Operations 2011). In the implementation of DRASTIC model, it has been modified it by adding a parameter corresponding to the anthropogenic factors, i.e., land use and land cover parameter and named it as DRASTICL. The parameter accounts for the distribution of land use and its role in groundwater contamination. The equation for estimation of vulnerability index for Modified DRASTIC model is as follows:

$$V\_\text{Index} = \sum_{i=1}^{8} W_i R_i,$$ (3.1)

where W and R stand for weights and ratings for eight parameters of Modified DRASTIC.

Estimation of Hydrogeological Parameters for Modified DRASTIC

Some of the classified hydrogeological parameters are shown in Figure 3.3.

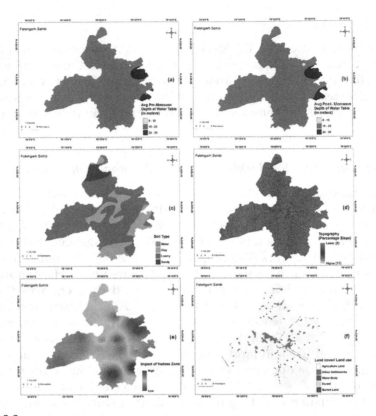

FIGURE 3.3
(a) Pre-monsoon depth of water table, (b) post-monsoon depth of water table, (c) soil map,
(d) topography map, (e) vadose zone impact profile, and (f) LULC map.

Depth of Water Table

Water table fluctuation is observed in two seasons, i.e., pre-monsoon and post-monsoon. As per the record of 5 years from 2008 to 2013, the pre-monsoon average depth of water table varies from 14.81 meter below ground level (mbgl) to 29.5 mbgl, and the post-monsoon average depth of water table varies from 14.45 to 30.1 mbgl in Fatehgarh Sahib (Kumar et al. 2016). The pre-monsoon and post-monsoon depth of water table map is shown in Figure 3.3a and b). The depth of water table has been classified into three classes with ratings corresponding to 5 (depth being 9–15 mbgl), 3 (depth being 15–23 mbgl) and 2 (depth being 23–30 mbgl). It is evident from Figure 3.3a and b that major portions of Fatehgarh Sahib district have the depth of water table in the range of 15–23 mbgl. There is not much significant variation in pre- and post-monsoon seasons in Fatehgarh Sahib district due to over consumption of groundwater.

Net Recharge

The average annual net recharge in Fatehgarh Sahib district is more than 250 mm in all the five blocks (Kumar et al. 2016). Therefore, it has been assigned a uniform rating of 9.

Aquifer Type

The whole of Fatehgarh Sahib district is a single unit of alluvial sediments, thereby assigning a uniform rating of 8 to aquifer type parameter. The aquifer type has a very significant influence on the groundwater contamination owing to its characteristics related to hydraulic conductivity and porosity of sediments.

Soil Type

The top 1–2 m surface and subsurface layers from the earth's surface are analyzed to fix the soil type. The major portions of Fatehgarh Sahib district is covered with loam, and the parameter soil type has been used as general loam type in the earlier attempt of groundwater vulnerability assessment as discussed in Chapter 2 (Kumar et al. 2016). Though loam is largely prevalent in Fatehgarh Sahib, but there are small segments which confirm the presence of clay and sandy as surface layers. Soil map has been gathered from NBSS&LUP (Indian National Bureau of Soil Survey and land Use Planning) at 1: 2, 50, 000 scale for the study. Figure 3.3c shows the soil profile of Fatehgarh Sahib district.

Topography

Fatehgarh Sahib district is primarily a flat region. Shuttle Radar Topography Mission (SRTM) 90 data has been used for the topography map in the earlier exercise of groundwater contamination assessment in Chapter 2 (Kumar et al. 2016). The slope values were generated with the SRTM 90 version 4 by using the spatial analyst software. Here, in the case of Modified DRASTIC model, the slope map has been prepared from CartoDEM version 3 in spatial analyst software. CartoDEM version 3 has better accuracy than STRM 90 data. This DEM (digital elevation model) was generated and downloaded from Bhuvan geo portal of National Remote Sensing Centre (ISRO), Government of India, Hyderabad, India. CartoDEM has stated accuracy of 8 m in vertical and 15 m in horizontal (Muralikrishnan et al. 2011; Muralikrishnan et al. 2013) as compared to SRTM 90 m version 4 with absolute horizontal and vertical accuracies better than 20 and 16 m, respectively, with 90% confidence intervals (Jarvis et al. 2008). The topography map of Fatehgarh Sahib district is shown in Figure 3.3d. It is primarily a flat region with percentage slope being the range of 0–2. The three classes of percentage slope (0–2, 2–6, and 6–12) have been assigned ratings of 10, 9, and 5, respectively.

Impact of Vadose Zone

Lithological structures below the earth surface are very important aspect of groundwater contamination. Drilling data till 60 m depth have been considered in the study area. There are total 24 locations across the five blocks of Fatehgarh Sahib from where drilling data have been taken for the estimation of impact of vadose zone below the earth surface. Lithology corresponding to each location is elaborated in this section.

- Khamanon: In block Khamanon, three villages—Chandiala, Kalewal, and Burj have been considered. The lithological features of Khamanon are shown in Figure 3.4. It can be noticed that in all the three villages, top surface is filled with clay and after 5–6 m, sand starts occurring. The size of the sand grains vary from fine to medium. In Kalewal village, there have been few instances where hard sticky clay has been found till 38 m.

- Bassi Pathana: Five villages under Bassi Pathana block are Jai Singhwala, Nandpur Kalod, Ladpur, Fatehgarh Sahib, and Talwara as shown in Figure 3.5. In Jai Singhwala, clay has been found till 19 m. Thereafter, it is fine to medium sand which is dominant in the layers. Nandpur Kalod has a rather simpler lithology in the sense

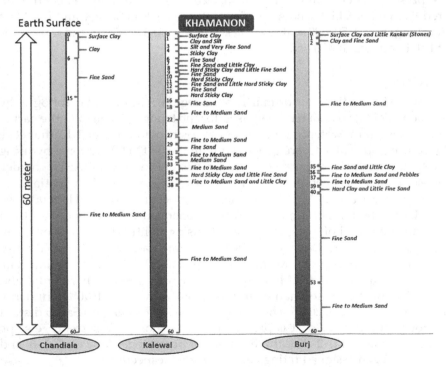

FIGURE 3.4
Lithological columns for wells of Khamanon block.

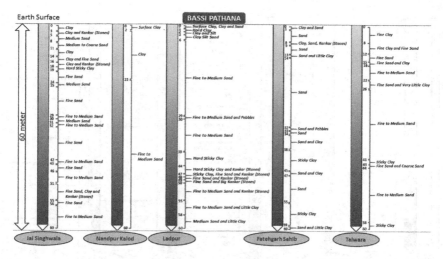

FIGURE 3.5
Lithological columns for wells of Bassi Pathana block.

that it is very regular. Clay has been found up to 23 m and the rest of the depth till 60 m is filled with fine to medium sand. Ladpur, Fatehgarh Sahib and Talwara have very irregular geometric lithology because there have been many occurrences of clay mixed with sand and it goes up to 60 m.

- Khera: Beru Majri, Timberpur, Salempur, Khera, and Saido Majra are the villages in Khera block as shown in Figure 3.6, where drilling has been done with a view to study the lithological structure of Khera block. In Beru Majri, major portions along the depth of 60 m was found to be rife with clay. There was small occurrence of fine to medium sized sand from 42 to 47 m. Timberpur is also dominated by clay. Clay has been found to be present till 38 m of depth. Thereafter, it is fine to medium sand which has been found from 38 to 51 m, and then rest of the depth till 60 m is full of clay. Salempur has an alternate presence of clay and sand with clay playing a major role. Khera village is full of clay till 23 m, and the rest of the depth till 60 m is found to be enriched with fine to medium sand. Saido Majra village is found to be dominated by both clay and sand till 45 m, and from 45 to 53 m, it is fine to medium sand which are present. Thereafter, it has been noticed that as the drilling went on, very little concentrated sandstones started occurring from 53 to 59 m.

- Sirhind: Sirhind is a bigger block of Fatehgarh Sahib district. The villages which have been considered under this block are Sirhind, Alipur Sodhiyan, Khare, Mianpur, Tehalpur, and Balpur as shown in Figure 3.7. Sirhind is full of clay mixed with sand till 60 m of depth.

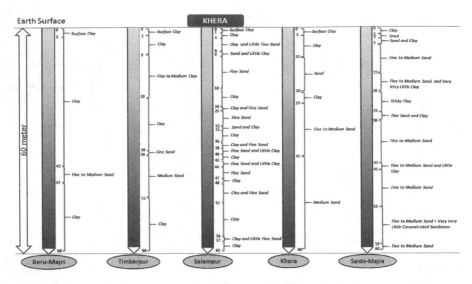

FIGURE 3.6
Lithological columns for wells of Khera block.

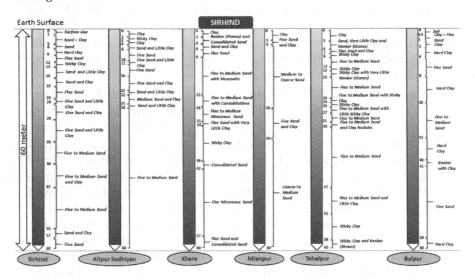

FIGURE 3.7
Lithological columns for wells of Sirhind block.

In Alipur Sodhiyan, half of the depth till 36 m are filled with clay and slight occurrences of sand. From 26 to 60 m, it is fine to medium sized sand which describe the lithology of Alipur Sodhiyan. Khare has been a bit different from the rest of the villages under consideration in Sirhind block. Here, clay and sand have been found till 9 m, and thereafter fine to medium sized sand mixed with muscovite has been found till 22 m. From 22 to 26 m, it is the fine to medium sized sand with consolidation which are found. Thereafter, clay continuous till 39 m and from 39 to 42 m, a small occurrence of consolidated sand is found. From 42 to 57 m, a very fine micaceous sand is found. Mianpur region is mostly dominated by medium to coarse sand which are found from 4 to 26 m and from 34 to 60 m. Tehalpur has a rather very zig zag occurrence of clay, sand, and little stone, and the region is mostly dominated by the presence of sticky clay. Balpur village is also dominated by clay which are found to very much hard in nature till 60 m.

- Amloh: Amloh block is located in south-western side of Fatehgarh Sahib district, and villages under this block identified for the study of drill data are Lohar Majra, Shahpur, Amloh, Jhambala, and Bagga Kalan as shown in Figure 3.8. Lohar Majra has been found to be full of clay and sand with several occurrences of sandstone concretions from 45 to 60 m. Shahpur has a regular lithology of sand and clay with coarse pebbles occurring at the end of 60 m. Amloh region is largely dominated by the presence of fine to medium sand till 54 m with beginning of clay afterwards. Jhambala region is full of fine to medium sand mixed with sticky clay. Also, there have been instances of little stones from 47 to 48 m in Jhambala. Bagga Kalan has been found to be full of fine to medium sand with alternate occurrences of sticky to hard clay.

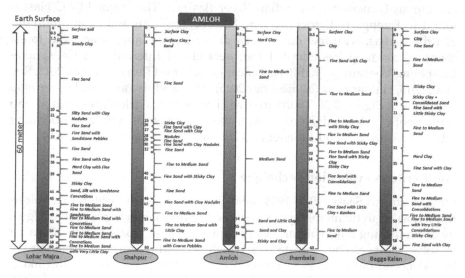

FIGURE 3.8
Lithological columns for wells of Amloh block.

The empirical formula to calculate the impact of vadose zone is given below:

$$\text{Vadose zone impact} = \left(\frac{1}{\text{total depth}}\right) \sum_{i=1}^{n} \text{depth}_{L_i} \times \text{rating}_{L_i}, \quad (3.2)$$

where depth_{L_i} = depth of the particular lithological unit such as sand, clay, stone, etc., rating_{L_i} = rating given to the particular lithological unit by Aller et al. (1987), total depth = depth of the water table from the earth's surface.

The resulting vadose zone profile is shown in Figure 3.3e indicating that major regions, shown in chartreuse color, fall into moderate potential zone from the perspective of impact of vadose zone.

Hydraulic Conductivity

Hydraulic conductivity regulates the movement of the contaminants in the groundwater. It is less than 5 m/day in Fatehgarh Sahib district. The lithological features described above in the assessment of vadose zone gives a reason for such a lower value, i.e., the major portions of lithological sample points are filled with clay along the line of depth with fine to medium sand up to 60 m.

Land Use/Land Cover

Fatehgarh Sahib district is largely an agrarian land. Land use land cover (LULC) map at medium resolution (~30 m) and 1:50000 scale have been taken from ISRO GBP project entitled "Land use/Land cover Dynamics and Impact of Human Dimensions in Indian River Basins." The major LULC classes found in Fatehgarh district are crop land or agricultural area and built-up or urban areas, with few minor area under barren land, water, and forest classes. This map has adopted the hierarchical International Geosphere-Biosphere Programme (IGBP) classification scheme (Tomaselli et al. 2013; Roy et al. 2015; Halder et al. 2016; Sikder 2016). The land use and land cover map is shown in Figure 3.3f. There are five classes, viz., agriculture land, urban settlements, water body, forest cover, and barren lands with corresponding ratings of 10, 8, 5, 3, and 1 (Sener and Davraz 2013; Sahoo et al. 2016a,b).

Multi-criteria Evaluation Technique

Mathematician Saaty (1977, 1988, 1990) proposed a multi-criteria technique known as analytic hierarchy process for solving/achieving a problem/goal by decomposing it into hierarchy of subproblems/subgoals, which are easier to solve/achieve. The main factors which serve the process of making the best decision out of several choices are importance, preference, and likelihood (Saaty 1977, 1988, 1990). People use the above three things to make any decision for a particular situation. AHP technique has found a very wider

acceptance in solving many complex problems such as conflict resolve, resource allocation, and any such issue which has many criteria for different alternatives (Vaidya and Kumar 2006). As already emphasized, fundamental DRASTIC model has one limitation in the sense that its ratings and weights assignments don't have any scientific backing and is based on human subjective opinion (Delphi network consensus). AHP tries to remove the subjectivity of Delphi network technique employed in DRASTIC model (Aller et al. 1987) by two ways:
    For any goal,

1. Subjective and comparative evaluation of qualitative and quantitative prospect of criteria leading to different alternatives.
2. Estimation and removal of inconsistencies in subjective evaluation.

AHP decomposes a problem or a goal in several criteria, sub criteria, and various alternatives and forms a hierarchy from goal to alternative. The relative weights of various criteria and alternatives are derived by forming a pairwise comparison matrix having relative judgment scores and calculation of normalized principal Eigen vectors. Further, the consistency of relatives weights are achieved by estimation of consistency indices and consistency ratios. If there is any inconsistency in the weights, a revision is made to assign the judgment values. In this study, a scale of 1–9 have used with different meanings to different levels of scale as shown below in Table 3.2 (Saaty 2008).
    The hierarchy of the goal, criteria, and subcriteria for the implementation of AHP-DRASTICL is shown in Figure 3.9.

TABLE 3.2

Saaty's Scale of Importance for Pairwise Comparison of Criteria

S. N.	Importance Scale	Meaning	Remarks
1	1	Equal	Experience and judgment rate both the criteria equally
2	3	Slightly important	Experience and judgment slightly recommend one criterion over other criterion
3	5	Moderately important	Experience and judgment moderately recommend one criterion over other criterion
4	7	Strongly important	Experience and judgment strongly recommend one criterion over other criterion
5	9	Absolutely important	Experience and judgment absolutely recommend one criterion over other criterion
6	2,4,6,8	Intermediate values between adjacent scales	Trade-off situations

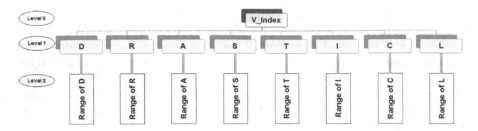

FIGURE 3.9
Hierarchy of goal, criteria, and subcriteria for AHP-DRASTICL.

Comparison Matrix and Consistency Ratio

Comparison matrices have been formed by the relative judgment of various criteria and subcriteria with respect to a particular goal. The ranking of the various criteria and subcriteria are decided by the estimation of the principal Eigen value of the matrix. In order to assess the consistency of the relative judgment to various criteria and subcriteria, consistency index is calculated as follows:

$$CI = (\lambda\_max - 1) / (n - 1),\qquad(3.3)$$

where λ_max = principal Eigen value, n = size of comparison matrix.

Further, consistency ratio is estimated, and if it is less than or equal to 10%, then it is acceptable. Otherwise, a revision of the relative judgment of the criteria and subcriteria is done. The consistency ratio is calculated as follows:

$$CR = CI / RCI,\qquad(3.4)$$

where RCI is standard random consistency index proposed by Saaty (2000) and Alonso and Lamata (2006).

Level 0 in Figure 3.9 is the main goal to deduce the vulnerability index of groundwater contamination. The level 1 and level 2 in Figure 3.9 are the subgoals which decide the final weights of the DRASTICL parameters and the ratings of the of various classes/ranges of the DRASTICL parameters, respectively. The following sections describes comprehensive method of optimization of ratings and weights for all the parameters.

Paired Comparison Matrix Level 1 w.r.t. V_Index
(Optimization of Weights for Various Parameters)

At level 1 with respect to the goal V_Index (level 0), the weights of all the hydrogeological parameters are derived by assigning Saaty's scale of importance to the parameters and their relative comparison. The pairwise comparison matrix for weights assessment is given in Table 3.3. The logic of assigning a particular value/ranking in pairwise comparison is elaborated here.

TABLE 3.3

Pairwise Comparison Matrix of DRASTICL Parameters for Weights

DRASTICL	D	R	A	S	T	I	C	L	Normalized Principal Eigen Vector (Final Weights)
D	1	2	③	4	5	1	3	1	0.2113
R	1/2	1	2	3	4	1/2	2	1/2	0.1274
A	①/3	1/2	1	2	3	1/3	1	1/3	0.0773
S	1/4	1/3	1/2	1	3	1/4	1/2	1/4	0.0523
T	1/5	1/4	1/3	1/3	1	1/5	1/3	1/5	0.0318
I	1	2	3	4	5	1	3	1	0.2113
C	1/3	1/2	1	2	3	1/3	1	1/3	0.0773
L	1	2	3	4	5	1	3	1	0.2113

Principal Eigen Value = 8.1459.
CI = (8.1459 − 8)/7 = 0.02084.
CR = CI/RCI = 0.02084/1.41 = 0.01478 = 1.48% (acceptable as it is less than 10%).

For the Table 3.2, all the original weights (1–5) based on Delphi network technique is mapped linearly to the new scales of importance (1–9) based on Saaty's scale with a condition that new weight/scale should be an integer to reduce the mathematical calculation. As depth of water table (D) is having a weight of 5 originally (equivalent to new scale of 9 as per Saaty's scale) and aquifer type (A) is having a weight of 3 originally (equivalent to new scale of 7 as per Saaty's scale), in the cell pertaining to D to A (horizontal to vertical), a value of 3 (encircled value in Table 3.3) is put as D is slightly important than A as per Saaty's scale of importance (Table 3.2). Further, in the cell pertaining to A to D (horizontal to vertical), a value of 1/3 (encircled value in Table 3.3) is put because it is just the inverse of the earlier case. Similarly the entire cells of the matrix is filled.

Paired Comparison Matrix Level 2 w.r.t. DRASTICL Parameters
(Optimization of Ratings for Various Parameters)

Paired Comparison Matrix Level 2 w.r.t. Depth of Water Table There are eight subgoals at level 2. At level 2, with respect to the subgoal Depth of Water table at level 1, the ratings of ranges of depth of water table are derived by assigning Saaty's scale of importance and their relative comparison. The pairwise comparison matrix for ratings assessment for parameter depth of water table is given in Table 3.4.

Paired Comparison Matrix Level 2 w.r.t. Net Recharge At level 2 with respect to the subgoal Net Recharge, the ratings of ranges of net recharge are derived by assigning Saaty's scale of importance and their relative comparison. The pairwise comparison matrix for ratings assessment is given in Table 3.5.

TABLE 3.4

Pairwise Comparison Matrix of Ranges of Depth of Water Table for Ratings

Ranges of D (m)	0–1.5	1.5–4.5	4.5–9	9–15	15–23	23–30	More than 30	Normalized Principal Eigen Vector (Final Ratings)
0–1.5	1	2	3	5	7	8	9	0.35
1.5–4.5	1/2	1	4	4	6	7	8	0.29
4.5–9	1/3	1/4	1	3	5	6	7	0.16
9–15	1/5	1/4	1/3	1	3	4	6	0.09
15–23	1/7	1/6	1/5	1/3	1	3	5	0.05
23–30	1/8	1/7	1/6	1/4	1/3	1	3	0.03
More than 30	1/9	1/8	1/7	1/6	1/5	1/3	1	0.02

Principal Eigen Value = 7.6353.
CI = (7.6353 − 7)/6 = 0.10588.
CR = CI/RCI = 0.10588/1.32 = 0.08021 = 8.02% (acceptable as it is less than 10%).

TABLE 3.5

Pairwise Comparison Matrix of Ranges of Net Recharge for Ratings

Range of R (mm)	0–50	50–100	100–175	175–250	More than 250	Normalized Principal Eigen Vector (Final Ratings)
0–50	1	1/3	1/6	1/8	1/9	0.0327
50–100	3	1	1/3	1/5	1/6	0.0685
100–175	6	3	1	1/3	1/3	0.1565
175–250	8	5	3	1	1/2	0.3109
More than 250	9	6	3	2	1	0.4329

Principal Eigen Value = 5.152.
CI = (5.152 − 5)/4 = 0.038.
CR = CI/RCI = 0.038/1.12 = 0.0339 = 3.39% (acceptable as it is less than 10%).

Paired Comparison Matrix Level 2 w.r.t. Aquifer Type   At level 2 with respect to the subgoal Aquifer Type, the ratings of classes of aquifer type are derived by assigning Saaty's scale of importance and their relative comparison. The pairwise comparison matrix for ratings assessment is given in Table 3.6.

Paired Comparison Matrix Level 2 w.r.t. Soil Type   At level 2 with respect to the subgoal Soil Type, the ratings of classes of soil type are derived by assigning Saaty's scale of importance and their relative comparison. The pairwise comparison matrix for ratings assessment is given in Table 3.7.

TABLE 3.6

Pairwise Comparison Matrix of Classes of Aquifer Type for Ratings

Aquifer Type	Massive Shale	Metamorphic/ Igneous	Weathered Metamorphic/ Igneous/Thin Bedded Sandstone, Limestone	Shale Sequences	Massive Sandstone	Massive Limestone	Sand and Gravel	Basalt	Karst Limestone	Normalized Principal Eigen Vector (Final Ratings)
Massive shale	1	1/2	1/4	1/5	1/5	1/6	1/6	1/6	1/7	0.021
Metamorphic/ igneous	2	1	1/3	1/4	1/4	1/5	1/5	1/5	1/6	0.028
Weathered metamorphic/ igneous/thin bedded sandstone, limestone	4	3	1	1/3	1/3	1/4	1/4	1/4	1/5	0.047
Shale sequences	5	4	3	1	1	1/3	1/3	1/3	1/3	0.081
Massive sandstone	5	4	3	1	1	1/3	1/3	1/3	1/3	0.081
Massive limestone	6	5	4	3	3	1	1	1/2	1/2	0.155
Sand and gravel	6	5	4	3	3	1	1	1/2	1/2	0.155
Basalt	6	5	4	3	3	2	2	1	1	0.211
Karst limestone	7	6	5	3	3	2	2	1	1	0.221

Principal Eigen Value = 9.4593.

CI = (9.4593 − 9)/8 = 0.0574.

CR = CI/RCI = 0.0574/1.45 = 0.0396 = 3.96% (acceptable as it is less than 10%).

TABLE 3.7

Pairwise Comparison Matrix of Classes of Soil Type for Ratings

Soil Type	Thin or Absent	Gravel	Sand	Peat	Shrinking/ Aggregated Clay	Sandy Loam	Loam	Silty Loam	Clay Loam	Muck	Non-shrinking/ Non-aggregated clay	Normalized Principal Eigen Vector (Final Ratings)
Thin or absent	1	1	2	2	3	4	5	6	7	8	9	0.211
Gravel	1	1	2	2	3	4	5	6	7	8	9	0.211
Sand	1/2	1/2	1	2	2	3	4	5	6	7	8	0.153
Peat	1/2	1/2	1/2	1	2	2	3	4	5	6	7	0.118
Shrinking/ aggregated clay	1/3	1/3	1/2	1/2	1	2	3	4	5	6	7	0.98
Sandy loam	1/4	1/5	1/3	1/2	1/2	1	2	3	4	5	6	0.070
Loam	1/5	1/5	1/4	1/3	1/3	1/2	1	2	3	4	5	0.049
Silty loam	1/6	1/6	1/5	1/4	1/4	1/3	1/2	1	2	3	4	0.034
Clay loam	1/7	1/7	1/6	1/5	1/5	1/4	1/3	1/2	1	2	3	0.025
Muck	1/8	1/8	1/7	1/6	1/6	1/5	1/4	1/3	1/2	1	2	0.018
Non-shrinking/ non-aggregated clay	1/9	1/9	1/8	1/7	1/7	1/6	1/5	1/4	1/3	1/2	1	0.014

Principal Eigen Value = 11.4864.

CI = (11.4864 − 11)/10 = 0.04864.

CR = CI/RCI = 0.04864/1.51 = 0.03221 = 3.22% (acceptable as it is less than 10%).

TABLE 3.8

Pairwise Comparison Matrix of Ranges of Topography for Ratings

Topography	0–2	2–6	6–12	12–18	More than 18	Normalized Principal Eigen Vector (Final Ratings)
0–2	1	3	5	7	9	0.513
2–6	1/3	1	3	5	7	0.262
6–12	1/5	1/3	1	3	5	0.129
12–18	1/7	1/5	1/3	1	3	0.063
More than 18	1/9	1/7	1/5	1/3	1	0.033

Principal Eigen Value = 5.2375.
CI = (5.2375 − 5)/4 = 0.05937.
CR = CI/RCI = 0.05937/1.12 = 0.05301 = 5.30% (acceptable as it is less than 10%).

Paired Comparison Matrix Level 2 w.r.t. Topography At level 2 with respect to the subgoal Topography, the ratings of ranges of topography are derived by assigning Saaty's scale of importance and their relative comparison. The pairwise comparison matrix for ratings assessment is given in Table 3.8.

Paired Comparison Matrix Level 2 w.r.t. Impact of Vadose Zone At level 2 with respect to the subgoal Impact of Vadose Zone, the ratings of vadose zones are derived by assigning Saaty's scale of importance and their relative comparison. The pairwise comparison matrix for ratings assessment is given in Table 3.9.

Paired Comparison Matrix Level 2 w.r.t. Hydraulic Conductivity At level 2 with respect to the subgoal Hydraulic Conductivity, the ratings of ranges of hydraulic conductivity are derived by assigning Saaty's scale of importance and their relative comparison. The pairwise comparison matrix for ratings assessment is given in Table 3.10.

Paired Comparison Matrix Level 2 w.r.t. LULC At level 2 with respect to the subgoal Land use/Land cover, the ratings of classes of land use/land cover are derived by assigning Saaty's scale of importance and their relative comparison. The pairwise comparison matrix for ratings assessment is given in Table 3.11.

TABLE 3.9

Pairwise Comparison Matrix of Classes of Vadose Zones for Ratings

Vadose Zone	Silt/ Clay	Shale	Lime- stone	Sand- stone	Bedded Limestone, Sandstone	Sand and Gravel with Significant Silt and Clay	Metamorphic/ Igneous	Sand and Gravel	Basalt	Karst Lime- stone	Normalized Principal Eigen Vector (Final Ratings)
Silt/clay	1	1	1	1/3	1/3	1/3	1/4	1/5	1/6	1/6	0.0275
Shale	1	1	1	1/3	1/3	1/3	1/4	1/5	1/6	1/6	0.0275
Limestone	1	1	1	1/3	1/3	1/3	1/4	1/5	1/6	1/6	0.0275
Sandstone	3	3	3	1	1/2	1/2	1/2	1/3	1/4	1/4	0.0578
Bedded limestone, sandstone	3	3	3	2	1	1/2	1/2	1/3	1/4	1/4	0.0667
Sand and gravel with significant silt and clay	3	3	3	2	2	1	1/2	1/3	1/4	1/4	0.0768
Metamorphic/ igneous	4	4	4	2	2	2	1	1/2	1/2	1/3	0.1072
Sand and gravel	5	5	5	3	3	3	2	1	1/2	1/2	0.1564
Basalt	6	6	6	4	4	4	2	2	1	1	0.2211
Karst limestone	6	6	6	4	4	4	3	2	1	1	0.2315

Principal Eigen Value = 10.3067.

CI = (10.3067 − 10)/9 = 0.0341.

CR = CI/RCI = 0.0341/1.49 = 0.0228 = 2.29% (acceptable as it is less than 10%).

TABLE 3.10

Pairwise Comparison Matrix of Ranges of Hydraulic Conductivity for Ratings

Hydraulic Conductivity	0.04074–4.074	4.074–12.222	12.222–28.518	28.518–40.74	40.74–81.48	More than 81.48	Normalized Principal Eigen Vector (Final Ratings)
0.04074–4.074	1	1/3	1/4	1/6	1/7	1/8	0.0274
4.074–12.222	3	1	1/3	1/5	1/6	1/7	0.0443
12.222–28.518	4	3	1	1/3	1/5	1/6	0.0759
28.518–40.74	6	5	3	1	1/3	1/5	0.1437
40.74–81.48	7	6	5	3	1	1/3	0.2580
More than 81.48	8	7	6	5	3	1	0.4506

Principal Eigen Value = 6.4987.
CI = (6.4987 − 6)/5 = 0.09974.
CR = CI/RCI = 0.09974/1.24 = 0.08043 = 8.043% (acceptable as it is less than 10%).

TABLE 3.11

Pairwise Comparison Matrix of Classes of Land Use/Land Cover

LULC	Agriculture Land	Settlements	Water Bodies	Forest Land	Barren Land/ Wastelands	Normalized Principal Eigen Vector (Final Ratings)
Agriculture land	1	3	5	7	8	0.5103
Settlements	1/3	1	3	5	7	0.2641
Water bodies	1/5	1/3	1	3	5	0.1306
Forest land	1/7	1/5	1/3	1	2	0.0579
Barren land/ wastelands	1/8	1/7	1/5	1/2	1	0.0372

Principal Eigen Value = 5.2093.
CI = (5.2093 − 5)/4 = 0.0523.
CR = CI/RCI = 0.0523/1.12 = 0.04671 = 4.672% (acceptable as it is less than 10%).

The ratings and weights derived after multi-criteria evaluation technique are given in Table 3.12 along with the fundamental ratings and weights given by Aller et al. (1987).

Based on the ratings of AHP-DRASTICL corresponding to Modified DRASTIC values given in Table 3.12, depth of water table for pre-monsoon and post-monsoon, soil type, percentage slope, vadose zone profile, and land use and land cover parameters are reclassified.

TABLE 3.12

Weights and Ratings of DRASTIC, Modified DRASTIC, and AHP-DRASTICL

Parameters	Range/Types	DRASTIC Weights	DRASTIC Ratings	Modified DRASTIC Weights	Modified DRASTIC Ratings	AHP-DRASTICL Weights	AHP-DRASTICL Ratings
Depth to water table (m)	0–1.5	5	10	5	10	0.2113	0.35
	1.5–4.5		9		9		0.29
	4.5–9		7		7		0.16
	9–15		5		5		0.09
	15–23		3		3		0.05
	23–30		2		2		0.03
	More than 30		1		1		0.02
Net recharge (mm)	0–50	4	1	4	1	0.1274	0.0327
	50–100		3		3		0.0685
	100–175		6		6		0.1565
	175–250		8		8		0.3109
	More than 250		9		9		0.4329
Aquifer media	Massive shale	3	2	3	2	0.0773	0.021
	Metamorphic/igneous		3		3		0.028
	Weathered metamorphic/igneous/thin bedded sandstone, limestone		4		4		0.047
	Shale sequences		6		6		0.081
	Massive sandstone		6		6		0.081
	Massive limestone		8		8		0.155
	Sand and gravel		8		8		0.155
	Basalt		9		9		0.211
	Karst limestone		10		10		0.221
Soil type	Thin or absent	2	10	2	10	0.0523	0.211
	Gravel		10		10		0.211
	Sand		9		9		0.153
	Peat		8		8		0.118
	Shrinking or aggregated clay		7		7		0.98
	Sandy loam		6		6		0.070
	Loam		5		5		0.049

(Continued)

TABLE 3.12 (*Continued*)

Weights and Ratings of DRASTIC, Modified DRASTIC, and AHP-DRASTICL

		DRASTIC		Modified DRASTIC		AHP-DRASTICL	
Parameters	Range/Types	Weights	Ratings	Weights	Ratings	Weights	Ratings
	Silty loam		4		4		0.034
	Clay loam		3		3		0.025
	Muck		2		2		0.018
	Non-shrinking or non-aggregated clay		1		1		0.014
Topography (percent slope)	0–2	1	10	1	10	0.0318	0.513
	2–6		9		9		0.262
	6–12		5		5		0.129
	12–18		3		3		0.063
	More than 18		1		1		0.033
Impact of vadose zone	Silt/clay	5	3	5	3	0.2113	0.0275
	Shale		3		3		0.0275
	Limestone		3		3		0.0275
	Sandstone		6		6		0.0578
	Bedded limestone, sandstone		6		6		0.0667
	Sand and gravel with significant silt and clay		6		6		0.0768
	Metamorphic/igneous		7		7		0.1072
	Sand and gravel		8		8		0.1564
	Basalt		9		9		0.2211
	Karst limestone		10		10		0.2315
Hydraulic conductivity (m per day)	0.04074–4.074	3	1	3	1	0.0773	0.0274
	4.074–12.222		2		2		0.0443
	12.222–28.518		4		4		0.0759
	28.518–40.74		6		6		0.1437
	40.74–81.48		8		8		0.2580
	More than 81.48		10		10		0.4506
Land use & land cover	Agriculture land	—		5	10	0.2113	0.5103
	Settlements				8		0.2641
	Water bodies				5		0.1306
	Forest land				3		0.0579
	Barren land/wastelands				1		0.0372

Implementation of Modified DRASTIC Model and AHP-DRASTICL Model

Estimation of Vulnerability Index for Modified and AHP-DRASTICL Model

The vulnerability indices corresponding to Modified DRASTIC and AHP-DRASTICL are calculated using Equation (3.1) with respective ratings for features of parameters and weights for parameters as given in Table 3.12.

The groundwater vulnerability map corresponding to the Modified DRASTIC is given in Figure 3.10. It has four high vulnerable zones as indicated in Figure 3.10. The highly vulnerable zones 1, 2, and 4 are the same as in the case of groundwater vulnerability map generated by fundamental DRASTIC model as shown in Figure 2.8. Highly vulnerable zone 3 in Figure 3.10 has not been highlighted in the vulnerability map of fundamental DRASTIC model in Figure 2.8 (Kumar et al. 2016). This could be due to the fact that Modified DRASTIC has considered a very clearly classified soil map wherein highly vulnerable zone 3 is full of sand as shown in Figure 3.3c. The porosity of sand is high which causes the contaminants to infiltrate to the groundwater. Also, land use and land cover maps would have played their own role.

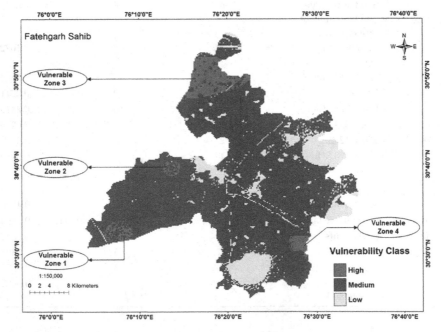

FIGURE 3.10
Vulnerability map for Fatehgarh Sahib using modified DRASTIC.

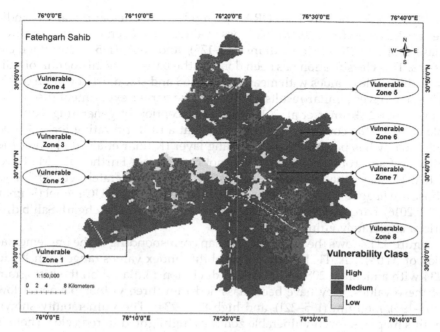

FIGURE 3.11
Vulnerability map for Fatehgarh Sahib using AHP-DRASTICL.

AHP-DRASTICL uses optimized weights and ratings for the hydrogeological parameters and the resultant vulnerability map is shown in Figure 3.11. There are eight highly vulnerable zones as indicated in Figure 3.11. The zones 1, 2, 7, and 8 in Figure 3.11 are highlighted as highly vulnerable zones which is similar to what is shown in Figure 2.8 (fundamental DRASTIC model) and Figure 3.10 (Modified DRASTIC model). Further, vulnerable zone 4 in Figure 3.11 is also flagged as highly vulnerable similar to what Modified DRASTIC has shown as vulnerable zone 3 in Figure 3.10. The new additions to the highly vulnerable segments are 3, 5, and 6 in Figure 3.11.

Comparative Analysis of Vulnerability

The assessment of the vulnerability of groundwater to contamination in Fatehgarh Sahib district using DRASTIC model was first done which highlighted four highly vulnerable zones (Kumar et al. 2016). The issue pertaining to the accuracy of vulnerability map generated when DRASTIC model is used and its effectiveness have also been raised (Kumar et al. 2016). This work is an attempt to resolve the fundamental issues of DRASTIC model. The vulnerability of groundwater in Fatehgarh Sahib district has been assessed via two modes—Modified DRASTIC and AHP-DRASTICL. The vulnerability map for Fatehgarh Sahib district using Modified DRASTIC model is shown in Figure 3.10. The Modified DRASTIC model generates vulnerability indices in the range from 107 to 182. These values have been classified in three classes of low (107–162), medium (162–175), and high (175–182) vulnerable zones. This classification has been done on the basis of the histogram of vulnerability index values with mean lying at 168 and standard deviation of 6.5. This map gives similar results as given in the earlier experiment discussed in Chapter 2 (Kumar et al. 2016) with an exception of generating vulnerable zone 3. This could be due to the fact that a uniform rating for soil type with loamy has been used as prevailing layer (Kumar et al. 2016). Also, the effects of anthropogenic factors was not considered. Further, the Modified DRASTIC flags the similar regions as lowly and moderately vulnerable as those are flagged by us in the case of Fundamental DRASTIC model (Kumar et al. 2016). Largely, Modified DRASTIC model declares Fatehgarh Sahib district a moderately vulnerable zone.

Figure 3.11 shows the vulnerability map corresponding to the implementation of AHP-DRASTICL. Here, vulnerability index values range from 105 to 234 with a mean at 208 and standard deviation 13. Based on the histogram of these values, they have been classified into three vulnerable zones low (105–195), medium (195–221), and high (221–234). The vulnerability shown in this map has seven vulnerable zones as highlighted in red color. There is a drastic change in this map because of the incorporation of anthropogenic factors as well as optimized ratings and weights as deduced by multi-criteria evaluation technique. This map, however, does highlight those areas as highly vulnerable which were earmarked as highly vulnerable by fundamental DRASTIC model (Kumar et al. 2016) and Modified DRASTIC model.

According to fundamental DRASTIC model (Kumar et al. 2016), 3.31%, 93.29%, and 3.4% area of Fatehgarh Sahib district are quantified as lowly, moderately, and highly vulnerable zones, respectively. In the experiment, it has been found that Modified DRASTIC earmarks 14.56%, 78.47%, and 6.97% area as low, medium, and high classes of vulnerability, respectively. Further, AHP-DRASTICL model suggests that 6.35%, 71.11%, and 22.54% area of the district are under low, medium, and high risk of contamination, respectively. Fundamental DRASTIC model highlighted very smaller region as highly vulnerable in comparison to Modified DRASTIC and AHP-DRASTICL because of the non-inclusion of anthropogenic factors such as improper waste disposal and excessive use of fertilizers and pesticides.

Validation

Validation of such kind of empirical studies is difficult in the sense that vulnerability maps generated by these empirical models, which take into account the physical parameters, can't be always very accurate due to the complex dynamics of groundwater. This can never replace the cumbersome manual on-site physical investigations for ground truth. But, DRASTIC, being a rapid regional assessment technique, can always give approximate vulnerability maps which can further be used for area-wise in-depth analysis. In order to validate the results, ten water quality parameters (Kalinski et al. 1994; Mogaji et al. 2014), viz., total alkalinity, pH, TDS, hardness, magnesium, sulphate, calcium, iron, fluoride, and nitrate have been used, available with us for 241 sample points spread uniformly across the study area.

Figure 3.12a–d shows show the distribution of water quality parameters— alkalinity, pH, TDS, and hardness of groundwater. Total alkalinity is measured by its capability of negating the acidic effects of groundwater. The main reason for the alkalinity is the bicarbonates formed in the soils through which water infiltrates. Bureau of Indian Standards (BIS) specifies a desirable limit of alkalinity up to 200 mg/L in drinking water and an acceptable limit up to 600 mg/L (Bureau of Indian Standards 2012). From Figure 3.12a, it can be seen that the alkalinity of groundwater of Fatehgarh Sahib district largely falls in the range of 300–500 mg/L. Vulnerable zones 1, 6, and 8 have relatively higher values of alkalinity in comparison to the rest of regions. The desirable pH range for drinking water by BIS is from 6.5 to 8.5 (Bureau of Indian Standards 2012; Chawla et al. 2015). Figure 3.12b shows that most of the regions of Fatehgarh Sahib district has pH range above 8. Also, it can be seen that all the vulnerable zones (1–8) fall in the higher range of pH values. The desirable and acceptable limits of TDS in

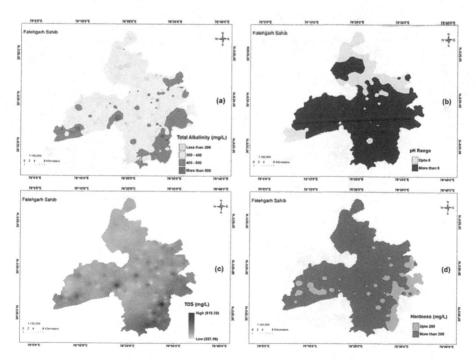

FIGURE 3.12
(a) Total alkalinity, (b) pH, (c) TDS, and (d) hardness.

drinking water are 500 and 2000 mg/L, respectively (Bureau of Indian Standards 2012). The TDS of groundwater in Fatehgarh Sahib district varies from 257 to 918 mg/L (Figure 3.12c) which is well within the acceptable limits with fewer regions (vulnerable zones 1, 2, 6, and 8) having values higher than the desirable value (Leal and Castillo 2003). The hardness distribution of groundwater samples shown in Figure 3.12d affirms the hard quality of water as major portions of Fatehgarh Sahib including the vulnerable zones 1–8 have hardness value above 200 mg/L which is the desirable limit set by BIS.

Further, the water quality parameters magnesium, sulphate, calcium, and iron for the study area were analyzed. The recommended desirable limits for magnesium, sulphate, calcium, and iron are up to 30, 200, 75, and 0.3 mg/L, respectively as stated by BIS (Bureau of Indian Standards 2012). Figure 3.13a shows that highly vulnerable zones 2, 3, 7, and 8 have magnesium more than 30 mg/L. Likewise, it can be seen from Figure 3.13b that vulnerable zone 8 has higher sulphate, otherwise sulphate is well within the desirable limits across the whole region. Similarly, calcium is also well within the desirable limits with several regions (highly vulnerable zones 1, 3, 4, 5, 6, and 7) having calcium up to the threshold value 75 mg/L (Figure 3.13c). Iron is well within the acceptable limits except in highly vulnerable zones 7 and 8 (Figure 3.13d).

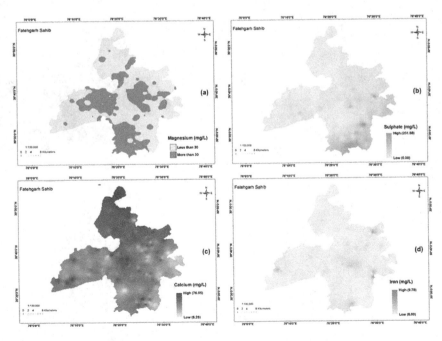

FIGURE 3.13
(a) Magnesium, (b) sulphate, (c) calcium, and (d) iron.

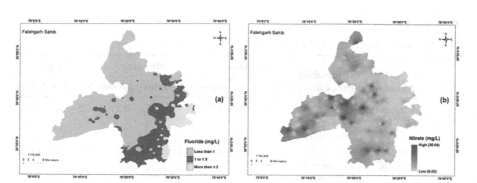

FIGURE 3.14
(a) Fluoride and (b) nitrate.

The last two water quality parameters are fluoride and nitrate. Fluoride is generally sparingly soluble in the groundwater and found in the form of calcium fluoride from the decomposition of rocks and soils and breakdown of atmospheric particles. Figure 3.14a shows that highly vulnerable zones 2, 3, 4, 5, 6, and 7 have fluoride in the range of 0.02–1.00 mg/L and vulnerable zone 8 has fluoride in the range from 1 to 2.2 mg/L as against the threshold of desirable limits for fluoride up to 1 mg/L as set by BIS. Nitrate is a very common contaminant in groundwater. The factors responsible for release of nitrate in groundwater are leaching of animal manure and fertilizers (Napolitano and Fabbri 1996; Thirumalaivasan et al. 2003; Huan et al. 2012). Figure 3.14b shows that nitrate is well within the desirable limits of 45 mg/L as specified by BIS with vulnerable zones 1, 2, and 8 having relatively higher values.

The vulnerable regions are validated after considering the cumulative effects of various water quality parameters due to complex and unpredictable dynamics associated with groundwater beneath earth's surface. Here, vulnerable zone 1 is dominated by the water quality parameters alkalinity, TDS, nitrate, calcium, and hardness. Vulnerable zones 2 and 6 are dominated by pH, TDS, alkalinity, fluoride, and hardness. Vulnerable zone 3 has higher cumulative effects of pH, alkalinity, nitrate, fluoride, calcium, magnesium, and hardness. Vulnerable zones 4 and 5 are influenced by pH, fluoride, alkalinity, calcium, and hardness. Vulnerable zone 7 has the severe effects of magnesium and iron along with higher effects of other water quality parameters. Similarly, vulnerable zone 8 has higher values for most of the water quality parameters such as pH, TDS, alkalinity, sulphate, nitrate, fluoride, and hardness.

Accuracy and Uncertainty Assessment

Table 3.13 presents the deviations in the vulnerability index and percentage areas pertaining to fundamental DRASTIC model and its two derivatives as part of accuracy assessment.

TABLE 3.13

Deviations in Vulnerability Index and Percentage Areas

S. N.	Models	Vulnerability Index Range	Low		Medium		High	
			V_index	%Area	V_index	%Area	V_index	%Area
1.	Fundamental DRASTIC	103–132 Mean = 119 SD = 4.03	100–110	3.31	110–125	93.29	125–140	3.4
2.	Modified DRASTIC	107–182 Mean = 168 SD = 6.5	107–162	14.56%	162–175	78.47%	175–182	6.97%
3.	AHP-DRASTICL	105–234 Mean = 208 SD = 13	105–195	6.35%	195–221	71.11%	221–234	22.54%

Conclusion

The present work is built upon the inherent lacuna of DRASTIC model wherein weights and ratings for various hydrogeological parameters are unscientifically assigned, along with the non-inclusion of anthropogenic factors. The consistent weights and ratings derived after relative evaluation by analytic hierarchy process of all the parameters have contributed in the more realistic assessment of vulnerability of groundwater to contamination. The accuracy of the same has been validated in the previous section. Also, it is worth noting here that anthropogenic factors such as agricultural practices, developmental works, and unscientific disposal of wastes all augment towards the groundwater contamination as highlighted by increased zones of study area under higher vulnerability by AHP-DRASTICL model.

Further, all the three models—DRASTIC, Modified DRASTIC, and AHP-DRASTICL suggest that major regions of Fatehgarh Sahib district are under moderate vulnerability from the contamination point of view. Such studies are very useful in groundwater monitoring and scientific land use planning for several purposes such agri-land usage, industrial land usage, and residential land usage. The accuracy of the geospatial groundwater vulnerability maps increases in the order of fundamental DRASTIC model, Modified DRASTIC model, and the AHP-DRASTICL model. The accuracy can further be enhanced by incorporating the physical contaminant transport model to assess the groundwater contamination propagation, but this is beyond the scope of this book.

References

Aller, L., T. Bannet et al. (1987). *DRASTIC: A Standardized System to Evaluate Ground Water Pollution Potential Using Hydrogeologic Settings*. Worthington, OH, National Water Well Association.

Alonso, J. A. and M. T. Lamata (2006). "Consistency in the analytic hierarchy process: A new approach." *International Journal of Uncertainty, Fuzziness and Knowledge-Based Systems* **14**(4): 445–459.

Arezoomand Omidi Langrudi, M., A. Khashei Siuki et al. (2016). "Evaluation of vulnerability of aquifers by improved fuzzy drastic method: Case study: Aastane Kochesfahan plain in Iran." *Ain Shams Engineering Journal* **7**(1): 11–20.

Ayele, G. T., S. S. Demessie et al. (2016). Multitemporal land use land cover change detection for the Batena Watershed, Rift Valley Lakes Basin, Ethiopia. *Landscape Dynamics, Soils and Hydrological Processes in Varied Climates*. M. A. Melesse and W. Abtew (Eds.), Cham, Switzerland, Springer International Publishing: pp. 51–72.

Bureau of Indian Standards (2012). *Drinking Water—Specification*. New Delhi, India, BIS, p. 11.

Chawla, P., P. Kumar et al. (2015). Prediction of pollution potential of Indian rivers using empirical equation consisting of water quality parameters. *Technological Innovation in ICT for Agriculture and Rural Development (TIAR)*. IEEE.

Chenini, I., A. Zghibi et al. (2015). "Hydrogeological investigations and groundwater vulnerability assessment and mapping for groundwater resource protection and management: State of the art and a case study." *Journal of African Earth Sciences* **109**: 11–26.

Directorate of Census Operations (2011). *District Census Handbook Fatehgarh Sahib*. Punjab, India, Ministry of Home Affairs, Government of India: p. 4.

Halder, S., S. K. Saha et al. (2016). "Investigating the impact of land-use land-cover change on Indian summer monsoon daily rainfall and temperature during 1951–2005 using a regional climate model." *Hydrology and Earth System Science* **20**(5): 1765–1784.

Huan, H., J. Wang et al. (2012). "Assessment and validation of groundwater vulnerability to nitrate based on a modified DRASTIC model: A case study in Jilin City of northeast China." *Science of the Total Environment* **440**: 14–23.

Huang, I. B., J. Keisler et al. (2011). "Multi-criteria decision analysis in environmental sciences: Ten years of applications and trends." *Science of the Total Environment* **409**(19): 3578–3594.

Iqbal, J., A. K. Gorai et al. (2015). "Development of GIS-based fuzzy pattern recognition model (modified DRASTIC model) for groundwater vulnerability to pollution assessment." *International Journal of Environmental Science and Technology* **12**(10): 3161–3174.

Jarvis, A., H. I. Reuter et al. (2008). "Hole-filled SRTM for the globe Version 4." CGIAR-CSI SRTM 90m Database from http://srtm.csi.cgiar.org.

Kalinski, R. J., W. E. Kelly et al. (1994). "Correlation between DRASTIC vulnerabilities and incidents of VOC contamination of municipal wells in Nebraska." *Groundwater* **32**(1): 31–34.

Kang, J., L. Zhao et al. (2016). "Groundwater vulnerability assessment based on modified DRASTIC model: A case study in Changli County, China." *Geocarto International* **32**: 749–758.

Kazakis, N. and K. S. Voudouris (2015). "Groundwater vulnerability and pollution risk assessment of porous aquifers to nitrate: Modifying the DRASTIC method using quantitative parameters." *Journal of Hydrology* **525**: 13–25.

Kumar, P., P. K. Thakur et al. (2016). "Assessment of the effectiveness of DRASTIC in predicting the vulnerability of groundwater to contamination: A case study from Fatehgarh Sahib district in Punjab, India." *Environmental Earth Sciences* **75**(10): 1–13.

Kumar, P., P. K. Thakur et al. (2018). "Groundwater: A regional resource and a regional governance." *Environment, Development and Sustainability* **20**: 1133–1151.

Lambin, E. F., B. L. Turner et al. (2001). "The causes of land-use and land-cover change: Moving beyond the myths." *Global Environmental Change* **11**(4): 261–269.

Leal, J. A. R. and R. R. Castillo (2003). "Aquifer vulnerability mapping in the Turbio river valley, Mexico: A validation study." *Geofísica Internacional* **42**(1): 141–156.

Mardani, A., A. Jusoh et al. (2015). "Multiple criteria decision-making techniques and their applications—A review of the literature from 2000 to 2014." *Economic Research-Ekonomska Istraživanja* **28**(1): 516–571.

Martín del Campo, M. A., M. V. Esteller et al. (2014). "Impacts of urbanization on groundwater hydrodynamics and hydrochemistry of the Toluca Valley aquifer (Mexico)." *Environmental Monitoring and Assessment* **186**(5): 2979–2999.

Mogaji, K. A., H. San Lim et al. (2014). Modeling groundwater vulnerability to pollution using optimized DRASTIC model. *IOP Conference Series: Earth and Environmental Science*. Bristol, UK, IOP Publishing.

Muralikrishnan, S., A. Pillai et al. (2013). "Validation of Indian national DEM from cartosat-1 data." *Journal of the Indian Society of Remote Sensing* **41**(1): 1–13.

Muralikrishnan, S., B. Narender et al. (2011). *Evaluation of Indian National DEM from Cartosat-1 Data*. Hyderabad, India, NRSC, p. 1.

Napolitano, P. and A. Fabbri (1996). "Single-parameter sensitivity analysis for aquifer vulnerability assessment using DRASTIC and SINTACS." *IAHS Publications-Series of Proceedings and Reports-Intern Assoc Hydrological Sciences* **235**: 559–566.

Nourqolipour, R., A. R. B. M. Shariff et al. (2016). "Predicting the effects of urban development on land transition and spatial patterns of land use in western Peninsular Malaysia." *Applied Spatial Analysis and Policy* **9**(1): 1–19.

Ouedraogo, I., P. Defourny et al. (2016). "Mapping the groundwater vulnerability for pollution at the Pan African scale." *Science of the Total Environment* **544**: 939–953.

Panagopoulos, G., A. Antonakos et al. (2006). "Optimization of the DRASTIC method for groundwater vulnerability assessment via the use of simple statistical methods and GIS." *Hydrogeology Journal* **14**(6): 894–911.

Rezaei, F., H. R. Safavi et al. (2013). "Groundwater vulnerability assessment using fuzzy logic: A case study in the Zayandehrood aquifers, Iran." *Environmental Management* **51**(1): 267–277.

Roy, P., A. Roy et al. (2015). "Development of decadal (1985–1995–2005) land use and land cover database for India." *Remote Sensing* **7**(3): 2401.

Saaty, T. L. (1977). "A scaling method for priorities in hierarchical structures." *Journal of Mathematical Psychology* **15**(3): 234–281.

Saaty, T. L. (1988). What is the analytic hierarchy process? *Mathematical Models for Decision Support.* G. Mitra, H. J. Greenberg, F. A. Lootsma, M. J. Rijkaert and H. J. Zimmermann (Eds.), Berlin, Germany, Springer, pp. 109–121.

Saaty, T. L. (1990). "How to make a decision: The analytic hierarchy process." *European Journal of Operational Research* **48**(1): 9–26.

Saaty, T. L. (2000). *Fundamentals of Decision Making and Priority Theory with the Analytic Hierarchy Process.* Pittsburgh, PA, RWS Publications.

Saaty, T. L. (2008). "Decision making with the analytic hierarchy process." *International Journal of Services Sciences* **1**(1): 83–98.

Sadat-Noori, M. and K. Ebrahimi (2015). "Groundwater vulnerability assessment in agricultural areas using a modified DRASTIC model." *Environmental Monitoring and Assessment* **188**(1): 1–18.

Sahoo, S., A. Dhar et al. (2016a). "Environmental vulnerability assessment using Grey Analytic Hierarchy Process based model." *Environmental Impact Assessment Review* **56**: 145–154.

Sahoo, S., A. Dhar et al. (2016b). "Index-based groundwater vulnerability mapping using quantitative parameters." *Environmental Earth Sciences* **75**(6): 1–13.

Saigal, S. K. (2007). *Ground Water Information Booklet Fatehgarh Sahib District, Punjab.* Chandigarh, India, Central Groundwater Board North Western Region.

Sener, E. and A. Davraz (2013). "Assessment of groundwater vulnerability based on a modified DRASTIC model, GIS and an analytic hierarchy process (AHP) method: The case of Egirdir Lake basin (Isparta, Turkey)." *Hydrogeology Journal* **21**(3): 701–714.

Shen, J., H. Lu et al. (2016). "Vulnerability assessment of urban ecosystems driven by water resources, human health and atmospheric environment." *Journal of Hydrology* **536**: 457–470.

Shouyu, C. and F. U. Guangtao (2003). "A DRASTIC-based fuzzy pattern recognition methodology for groundwater vulnerability evaluation." *Hydrological Sciences Journal* **48**(2): 211–220.

Sikder, I. U. (2016). "A variable precision rough set approach to knowledge discovery in land cover classification." *International Journal of Digital Earth* **9**: 1–18.

Singh, B. (1998). *Geoenvironmental Appraisal of Fatehgarh Sahib District, Punjab.* Kolkatta, India, Geological Survey of India.

Thirumalaivasan, D., M. Karmegam et al. (2003). "AHP-DRASTIC: Software for specific aquifer vulnerability assessment using DRASTIC model and GIS." *Environmental Modelling & Software* **18**(7): 645–656.

Tomaselli, V., P. Dimopoulos et al. (2013). "Translating land cover/land use classifications to habitat taxonomies for landscape monitoring: A Mediterranean assessment." *Landscape Ecology* **28**(5): 905–930.

Vaidya, O. S. and S. Kumar (2006). "Analytic hierarchy process: An overview of applications." *European Journal of Operational Research* **169**(1): 1–29.

Zhang, R., L. Pu et al. (2015). "Landscape ecological security response to land use change in the tidal flat reclamation zone, China." *Environmental Monitoring and Assessment* **188**(1): 1–10.

4

Groundwater Governance

Groundwater is freely accessible and this has led to the overconsumption and extreme exploitation of this resource over the period of time. The worth of groundwater can be realized by measuring the extent to which people across the world have increased their dependency over the groundwater for various purposes. The dependence over the groundwater has increased by four times in the last 50 years, especially in South Asia, Middle East, and North China. As a result, there is an urgent need for a groundwater governance framework for addressing the current and upcoming challenges associated with the groundwater resource. Groundwater development is very important from the perspectives of economic sustenance as it affects various aspects of human life in terms of agriculture, safe water supplies in rural and urban areas.

Groundwater governance framework is about formation of policies and the adoption of the best practices for tackling the issues related to the groundwater resources in a way to achieve the sustainable management of groundwater resources (Hellström et al. 2000; Almasri 2007; Kulkarni and Shankar 2009; Foster and Garduño 2013; Howard 2015; Pandey 2016). In the past 10 years, the governance and management of groundwater has caught the eyes of various renowned institutions and agencies such as World Bank (Foster et al. 2010) and National Ground Water Association (Shah et al. 2003; Shah 2014). The alarming concerns related to groundwater motivated the five organizations, viz., Global Environment Facility, World Bank, Food and Agriculture Organization of the United Nations, UNESCO-International Hydrological Programme (IHP), and International Association of Hydrogeologists to set up a full-fledged joint groundwater governance project in 2011, and these organizations came out with certain initiatives for the betterment of the groundwater in the form a comprehensive groundwater governance framework [UNESCO's International Hydrological Programme (UNESCO-IHP) 2015].

Water stress is affecting approximately 2 billion people across the globe as of now. United Nations came up with 17 sustainable developmental goals in the year 2015, and one of the goals in that list is water security for

This chapter is based on **Prashant Kumar et al.**, Groundwater: A regional resource and a regional governance, Environment, Development and Sustainability, Springer, Volume 19, Feb 2017, Issue 5, Pages 1–19 **(Impact Factor- 1.01)**.

all (Tengberg 2015; Gupta and Vegelin 2016). Universal access to drinking water, sanitation, and cleanliness can't be accomplished without the advancement and legitimate administration of groundwater assets. Water security is critical to the survival of people and the planet and inherently it hinders the sustainability of the natural resources and economic development.

Groundwater is a global entity, but it is actually a local resource. Therefore, in order to understand the ground realities of the issues pertaining to the groundwater, local governing analysis of groundwater resources is very essential (Maleksaeidi et al. 2016). Further, the community participation is very much needed in promotion of the best practices regarding groundwater resources, waste management, and land use. The socio-ecological and political environment play a vital role in groundwater governance. There is no best way to govern the groundwater regime because of the involvement of several socio-ecological and political issues. Therefore, it would be quite prudent to state that groundwater governance is not only about groundwater and aquifers, but also the social systems, stage of economic evolution, and the society's political organization (Mukherji and Shah 2005; Shah 2005; Giordano 2009; Shah 2014; Kulkarni et al. 2015; Wezel and Weizenegger 2016).

As majority of drinking water comes from groundwater, it becomes utmost crucial to have a look at the present groundwater governing style to protect it further for the future generations. It is an attempt to present the effective groundwater governance at a regional scale using Fatehgarh Sahib district of Punjab. Fatehgarh Sahib is an agrarian land, already highlighted as an overexploited region. The major issues being faced by Fatehgarh Sahib district are the contamination and the alarming rate of decline in water table. The chapter presents the local situation of groundwater governance in Fatehgarh Sahib district and deliberates upon the remedial measures to be taken for the ongoing deteriorating condition of groundwater in Fatehgarh Sahib district.

Exploratory Survey of the Study Area

On the visit to Fatehgarh Sahib district, it was found that Fatehgarh Sahib district being a rural region, had largely been ignored for developmental works. Figure 4.1 is an indicative representation of discussions held with the village chief and several countryside people.

Upon discussions with the local residents, following reasons were zeroed in for the deteriorating conditions of groundwater in Fatehgarh Sahib district.

FIGURE 4.1
Discussions with village chief and people of Fatehgarh Sahib district.

Small Land Holding

Landholding varies from marginal to small scale. The average landholdings in Fatehgarh Sahib district is less than one hectare and farmers are dependent on groundwater for irrigation. Fragmentation of small landholdings and tiny land parcels have always been counterproductive, thereby discouraging farmers from adoption of agricultural innovations (Niroula and Thapa 2005). Further, cooperative farming is not practices by consolidating these small lands together because of the very fear of losing the ownership rights over the lands. These small landholdings has caused enormous increase in the installation of submersible pumps by every land owner. The number of tube wells per acre of land is more than what it should be ideally.

Limited Power Supply

Though rural electrification has happened in the village on a reasonable scale, but power is supplied only for 6 and 7 hours. As a result of this, people run many submersible pumps located in the farmland simultaneously in the short duration, causing the groundwater level decline severely. The submersible pumps being used are also of very high capacity, viz., minimum 15 horsepower in order to pump out the water lying at a greater depth from the earth surface.

Hand Pump versus Submersible Pumps

Earlier, most of the households had hand pumps for the domestic need of water. In the last decade, every household has got a submersible pump installed in their house. This causes severe wastage of water because a submersible pump yields more water than what is needed for domestic works as uttered by the village people. This is also due to the fact that there is no storage facilities in most of the households and they run the submersible pump every time they need water, subjected to the availability of the power supply.

Routine Cultivation

Agriculture is the main source of living in Fatehgarh Sahib district. Mainly, paddy and sugarcane are cultivated depending upon the seasons throughout the year. These are water-intensive crops resulting in extensive irrigation, thereby lowering down the level of groundwater. One of the most interesting part of the agricultural practice in Fatehgarh Sahib district is that not a single farmer produces pulse or vegetables and the reason cited is the commercial inviability of these crops. The minimum support price for pulse is very less and it's not at all an economically beneficial crop for the farmers. On further scouting, it was revealed that farmers didn't produce vegetables because vegetables had been found to be vulnerable to the attack by blue bulls and it had caused extreme damages.

Geopolitical Reasons

Fatehgarh Sahib district has no major river and the entire agriculture is dependent on groundwater and canals for irrigation. Though there is Sutlez-Yamuna Link, but no water has been supplied in that link because of the ongoing fight between Punjab and Haryana resulting in the geopolitical causes for such initiatives.

Water Resources

Fatehgarh Sahib district has an extensive network of organized water system consisting of canals, distributaries, and minors as shown in Figure 2.2.

Surface Water

There are no major rivers in Fatehgarh Sahib district. The main sources of irrigation are tube wells and canals (Baldev 1998). The major canals passing through Fatehgarh Sahib are Bhakra canal, Sirhind canal, and Satluj-Yamuna Link canal as shown in Figure 2.2. Surface water has been discussed in detail in Section "GIS for Hydrological Investigations" under study area in Chapter 2.

Groundwater

Punjab, being an agrarian land, has been always at the front of the agricultural developments. Irrigation by groundwater in the districts of Punjab goes back to decade of sixties. Only 33% of the total irrigation was done by wells and tube wells before independence of India (Central Ground Water Board 2014). However, the area irrigated by groundwater has increased thereafter. Irrigation by groundwater in the Fatehgarh Sahib district is done mainly through shallow tube wells owned by farmers. Of the total geographic area of Fatehgarh Sahib district (1174 km^2) (Kumar et al. 2016), the area irrigated by groundwater is 1020 km^2 and the number of tube wells being utilized in Fatehgarh Sahib district is 35932 as of year 2007 (Saigal 2007).

Groundwater Fluctuation

Groundwater fluctuation is not significant in Fatehgarh Sahib district over the change of seasons as far as pre-monsoon and post-monsoon seasons are concerned. In time span of 5 years from year 2008 to 2013, the average pre-monsoon (June) depth of water table varies from 14.81 to 29.5 meters below ground level (mbgl) and the average post-monsoon (October) depth of water table varies from 14.45 to 30.1 mbgl (Kumar et al. 2016). Most of the regions of Fatehgarh Sahib district fall in the range of depth of water from 15 to 23 mbgl. The overexploited regions, corresponding to the depth of water table range 23–30 mbgl, of the district lie in the blocks Khera and Bassi Pathana. Post-monsoon situation almost remains same due to heavy surface runoff and high rate of evaporation (Kumar et al. 2016). Fatehgarh Sahib is already declared as an overexploited district of Punjab due to excessive tube wells. The rate of decline of water table has been found to be around 20–90 cm per year in the last 10 years (Saigal 2007; Kumar et al. 2016).

Groundwater Quality and Contamination

Most of the water quality parameters are within the permissible limits for drinking water in case of shallow aquifers found in Fatehgarh Sahib district. Deeper aquifers have somewhat alkaline water with pH values ranging from 8.22 to 8.60 (Saigal 2007). Fatehgarh Sahib district encompasses the steel town of northern India, i.e., Mandi Gobindgarh. Chemical analysis of urban area of Mandi Gobindgarh states that ground water from shallow and deep aquifers is somewhat good for various purposes such as household activities and irrigation. Industrial pollution is a big issue here and few shallow aquifers are contaminated with heavy metal ions such as Cu, Zn, Pb, and Fe due to disposal of industrial wastes. The main reason for the groundwater contamination in Fatehgarh Sahib can be attributed to the discharge from farming activities wherein excessive insecticide and pesticides are being used (Farooqi et al. 2009). Also, the industrial pollution due to industrial plants located in Mandi Gobindgarh plays a significant role in groundwater contamination. Waste disposal patterns also contribute in polluting the groundwater. There are two major causes for the groundwater contamination as given below.

Geogenic Causes

Geogenic causes of the groundwater refer to the naturally occurring higher concentrations of harmful elements like fluoride, arsenic, and uranium (Amini et al. 2008; Sikdar et al. 2013; Banerjee 2015; Chakraborty et al. 2015; Kumar et al. 2016; Saha and Sahu 2016). Geogenic contamination of the groundwater happens due to the geochemical properties of the aquifer medium, i.e., disintegration of contaminant in rock during water-rock face off or peculiar environmental conditions which brings the contaminants to occur in a roving state. Table 4.1 shows commonly found geogenic contaminants and its sources in Fatehgarh Sahib district.

Anthropogenic Causes

The anthropogenic contamination of groundwater refers to the human activities such as agriculture, urban settlements, industries, and plants, which assist the deterioration of groundwater quality (Barth 1998; Williams et al. 1998; Haase 2009; Rao et al. 2013; Eulenstein et al. 2016; Karkra et al. 2016; Zhu et al. 2016).

Fatehgarh Sahib is an agrarian land and consumes a lot of fertilizers. Over the period of time, farmers have been using a big amount of chemical fertilizers containing excessive nitrogen instead of organic fertilizers to get enhanced production which has caused soil and water pollution (Mittal et al. 2014). Nitrogen-based fertilizers discharge large amount of nitrates which accumulate in soil and reaches to groundwater through leaching (Datta et al. 1997;

TABLE 4.1

Geogenic Contaminants and Its Sources in Fatehgarh Sahib

S. No	Contaminants	Sources
1	Selenium	Selenium rich sediments as selenites & selenates
2	Arsenfic	Geothermal activities, weathering of rocks & minerals, herbicides, pesticides
3	Fluoride	Limestone, sandstone, granite, phosphatic fertilizers, disinfectants preservatives, insecticides
4	Nitrate	Nitrogen rich wastes buried on the land surface, fertilizers

Bukowski et al. 2001; Böhlke 2002; Tariq et al. 2004; Suthar et al. 2009; Nagarajan et al. 2010). Nitrate leaching is a serious problem in Fatehgarh Sahib district (Sharma et al. 2016). In rural areas, animal waste disposal also augments in nitrification of groundwater. Ultimately, all these contaminants are entering in the food chain.

Fatehgarh Sahib has well organized industrial units such as food products beverages, leather products, wood and paper products, metal products, and chemical products (Government of India, Ministry of MSME 2011). Disposal of untreated industrial effluents in the hydrological cycle causes severe groundwater contamination (Krishna and Govil 2004; Srinivasa Gowd et al. 2010). Apart from this, wastes resulted from manufacturing and processing units also pollute groundwater.

The urban pollution of groundwater is happening due to unplanned disposal of sewerage waste (Aulakh et al. 2009; White et al. 2016). Usage of cesspools or septic tanks cause contamination to groundwater (Bamaga et al. 2016). Storage tanks of petroleum products, acids can develop leakage and cause contamination to groundwater.

Groundwater Governance and Management

The two major aspects of groundwater are contamination and alarming rate of decline of groundwater level. Figure 4.2 proposes the groundwater governance framework as shown below.

Groundwater Contamination

Groundwater contamination is the result of both natural and anthropogenic activities. Excessive usage of fertilizers, pesticides, and unscientific disposal of industrial effluents and urban wastage play the most significant roles in

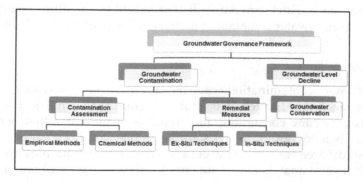

FIGURE 4.2
Groundwater governance framework.

the groundwater contamination. To govern the groundwater contamination, it is necessary to earmark the potential zones vulnerable to contamination and its remedial measures to counter the contamination.

Assessment of Contamination/Vulnerability to Contamination

Groundwater contamination assessment is mainly done by lab-based chemical methods, however, the vulnerability of the groundwater to contamination can be done by empirical methods. The empirical methods are rapid regional assessment methods which don't require rigorous field visits and generate the approximate groundwater vulnerability maps of the particular region. The lab-based chemical analysis is used as a conventional method to estimate the presence of contaminants in groundwater. This is more reliable and evidence based.

Empirical Methods

With the advancement of remotely sensed spatial and temporal data, assessment and monitoring of ground water resources have got immense applications of its usage for quick and comprehensive examination of groundwater contamination. Groundwater vulnerability maps resulting from empirical studies help in groundwater contamination assessment. Such studies can be augmented with the addition of climate projection models to see the effect of climatic conditions and the probable direction of propagation of contamination. Though such studies can never replace on-site visits, but these are very useful as rapid assessment tools for the estimation of groundwater vulnerability to contamination (Kumar et al. 2015). Various empirical models have been developed so far to estimate the vulnerability of groundwater contamination, and their accuracy have been assessed with lab-based chemical data (Gogu and Dassargues 2000; Kumar et al. 2015). These models use several hydro-geological parameters to build an empirical equation with an aim to assign a particular index to a region corresponding to the severity of contamination (Gogu and Dassargues 2000; Kumar et al. 2015). As per the study, Fatehgarh Sahib district falls in the medium vulnerability from the groundwater contamination perspective (Kumar et al. 2016).

Chemical Methods

Groundwater contamination can be assessed in laboratory through chemical analysis. Though it is more accurate than empirical methods, but it is very time consuming and rigorous in the sense that one needs to collect as many groundwater samples as one can, from various regions of the study area for a better accuracy. Groundwater samples are passed through several analytical instruments for the estimation of water quality parameters (Ramakrishnaiah, Sadashivaiah et al. 2009; Rasool et al. 2016; Thakur et al. 2016; Tziritis et al. 2016).

Remedial Measures

Groundwater contamination cannot be negated without the participation of both industrial sector and industries. The remedial measures include reasonable use of toxic materials, sewage water treatment and scientific disposal of industrial wastes, and reduced use of fertilizers and pesticides (Finizio and Villa 2002). Two classes of remedial measures are discussed below. The selection of remedial measures are dependent on two important factors, i.e., contaminant profile and aquifer profile.

Ex-Situ Techniques

De-watering the contaminated water from the polluted aquifer, treatment of the contaminated water through physio-chemical methods, and then watering the treated water back to the aquifer form the basis for *ex-situ* techniques. There are several *ex-situ* techniques which have been employed in the groundwater remediation globally. The most prevalent techniques are steam stripping (steam injection to the pumped out water and contaminant removal) (Kosusko et al. 1988; Davis 1998) and bioremediation (treatment of extracted groundwater with air under controlled conditions of moisture, pH, and heat) (Hatzinger et al. 2002). *Ex-situ* techniques are a bit costly as it involves pumping out the groundwater, but it maintains the uniformity of the treatment.

In-Situ Techniques

Here, extraction of the groundwater is not required. The physical, chemical, or biological methods are employed directly on groundwater within the aquifer (Kuppusamy et al. 2016). The most popular *in-situ* techniques are air sparging (injection of fresh air into subsurface saturated zone thereby converting hydrocarbons from dissolved to vapor phase) (Marley et al. 1992; McCray and Falta 1996; Rabideau et al. 1999), bioremediation (injection of oxygen to enhance biodegradation) (Minsker and Shoemaker 1998; Farhadian et al. 2008), and thermal treatment (increasing the temperature of groundwater zone to mobilize the contaminants for its destruction) (Heron et al. 2009; Baker et al. 2016). As per the land use and land cover analysis of Fatehgarh Sahib, the larger portions are agriculture land, bioremediation can be one of the most viable options to purify the groundwater.

Groundwater Level Decline

Fatehgarh Sahib district has been declared as an overexploited zone by Central Groundwater Board (Saigal 2007). The rate of decline is very significant in the range of 20–90 cm per year in the last 10–15 years (Kumar et al. 2016).

The reason for such a severe decline is excessive use of the tube wells for irrigation purpose. There are several groundwater conservation strategies which can be utilized to counter the declining level of groundwater.

Groundwater Conservation

The adoption of artificial recharge techniques can help maintain the required groundwater level in region where groundwater level is slumping due to overuse or where the aquifer is almost desaturated or where the quality of water is poor and no options for surface water source (Bouwer 2002; Ghayoumian et al. 2007; Aggarwal et al. 2009; Zektser et al. 2012; Kumar et al. 2016). Artificial recharge is a faster way of replenishment of groundwater. Three major artificial recharge techniques are surface recharge, subsurface recharge, and induced recharge. Surface techniques such as flooding, percolation tank, steam augmentation, and ditch and furrow system are predominantly used for unconfined aquifers where the rate of infiltration is higher. Recharge well, dug wells are part of subsurface recharge systems. The induced recharge technique is used in areas where the quality of surface water is not good. This technique makes a gradient in the aquifer system due to extraction of groundwater which subsequently gets filled with water percolated through soils from surface water sources (NGWA 1999). Rainwater harvesting and artificial recharge have been implemented in selected zones of Fatehgarh Sahib district. Recharge wells with filtration chambers have been built at these sites for effective augmentation of groundwater level. Artificial recharge with plant purification is very much preferable in areas where subsurface layers have sand and gravel (Balke and Zhu 2008).

Groundwater Governance

In order to provide a conducive environment for groundwater management, institutional set up is a necessary. Central Groundwater Board is one such institution under Ministry of Water Resources in Indian context. Despite its existence for so many decades, India needs to build up some effective institutions with proper laws and regulations to achieve sustainable developmental goal of water security at the regional level. The regulatory framework must emphasize primarily on three points, viz., groundwater is a public resource and one can't just drill as much as one desires, government licensing for groundwater extraction, and groundwater pollution control. As per the changing needs of water in a region for various domestic and industrial applications, the regulatory framework must account for the following regulations:

- A periodic assessment of water demands in response to the local/ global change and tenure-based groundwater abstraction licensing.
- Any rights pertaining to groundwater abstraction/usage detrimental to environment should be forfeited.

No matter how strong the laws are made for groundwater management, it is the implementation and enforcement that finally matter. All the stakeholders such as government officials, industrialists, and the local people, all are equally accountable for the execution of the regulations pertaining to the governance arrangements.

Groundwater Governance Framework

With community-based groundwater governance (as proposed in this study for Fatehagrh Sahib district) being at the helm of affairs for the effective groundwater management, it is very much significant to make sure that community-based activities gets the compulsory support from the central and state government agencies. The community awareness and participation is very important for groundwater governance chain, but the most crucial role has to be played by the government agencies to promote self-regulation. The state government, being a guardian, should enhance the interaction among various stakeholders such as local authorities, people, and supporting agencies. The regulatory bodies must enforce the guidelines at a local level for effective management of groundwater resources. The conceptual framework for groundwater governance on a regional scale in Fatehgarh Sahib district is shown in Figure 4.3.

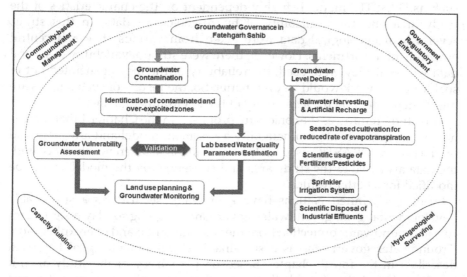

FIGURE 4.3
Groundwater governance framework in Fatehgarh Sahib.

Although UNESCO's International Hydrological Programme has developed a very consolidated roadmap for the global groundwater governance, the current study stands out as a case study or illustrations of how some of the initiatives of the International Hydrological Programme could be implemented at a regional scale considering the regional conditions and feasibility. Fatehgarh Sahib district will have serious groundwater issues in terms of both quality and quantity if a regional groundwater governance is not implemented very soon (Kumar et al. 2016). The global framework envisaged by the UNESCO's IHP intends to have its initiatives implemented by all the countries by 2030 to meet their targets of social and economic developments with a view to avoid irreversible damage to the aquifer system (UNESCO's International Hydrological Programme [UNESCO-IHP] 2015).

Conclusion and Future Prospects

Groundwater vulnerability assessment has emerged as a rapid regional assessment tool to segregate the region into various classes of severity of groundwater contamination. It doesn't involve the cumbersome job of water sample collections from many sample points from the study area thereby reducing the physical visit to the entire study area. The maps generated out of such empirical models like DRASTIC meet the very purpose of groundwater monitoring on a larger scale without involving too much resources. The application of empirical vulnerability assessment models such as DRASTIC model is highly dependent on the characteristics of the study area and the availability of hydro-geological data. In this study, several hydro-geological data were taken from various state and central government departments. However, there were still unavailability of several data which, if they were used, the reliability of the maps generated out of such research work would be even better. So, the results of such studies are highly dependent on the availability of the environmental data.

This study reinforces the conjecture proposed in the Chapter 1 that there is no generic groundwater vulnerability assessment model and any empirical model such as DRASTIC used for the study must be taken in coherence with the data available to the study area and subsequently the model should be modified for a better result.

Further, as United Nations have kept water security as a sustainable developmental goal, groundwater governance is going to play a significant role in groundwater protection from the contamination and overexploitation. Groundwater governance as a sustainable developmental goal has been described in the fourth chapter. Based on the survey of the study area, it has been found that though there are some geo-political reasons beyond control of an individual, still there are several factors/variables which, if assessed on an individual scale, can be controlled by the individual, and it will lead to a groundwater governance framework as a model framework. It is recommended to have a community-based groundwater management system, supplemented by demand management interventions, technical interventions, and policy enforcement.

This work has several future prospects. Generation of groundwater vulnerability maps using past hydro-geological data is static in nature. To bring the dynamism aspect, the estimation of the groundwater contamination propagation in several directions using climate change projection models can be very much useful. All the hydro-geological parameters under DRASTIC model in the current study belong to the last 10 years. The earth's climate is not static; it changes in response to both natural and anthropogenic drivers. Therefore, it would be interesting to see how the effect of precipitation and temperature rise through global futuristic models affect the groundwater contamination. This is a future prospect of this study. Such studies can also be done in association with physical contaminant transport models for even better groundwater vulnerability map development as part of future prospects. Any uncertainties associated with the groundwater vulnerability maps generated by DRASTIC or its derivative models can also be assessed in a comparative manner.

Since entire Punjab is facing groundwater problems, the work can be carried forward in future to assess the vulnerability of groundwater to other severely contaminated districts of Punjab such as Bhatinda, Ludhiana, and other regions of Malwa region of Punjab. This will help in creating a database of realistic groundwater vulnerability maps of all the districts of Punjab which can be used in groundwater monitoring and urban planning for the entire state. This will also help in building an integrated decision support system for efficient irrigation system.

References

Aggarwal, R., M. Kaushal et al. (2009). "Water resource management for sustainable agriculture in Punjab, India." *Water Science and Technology* **60**(11): 2905–2911.

Almasri, M. N. (2007). "Nitrate contamination of groundwater: A conceptual management framework." *Environmental Impact Assessment Review* **27**(3): 220–242.

Amini, M., K. C. Abbaspour et al. (2008). "Statistical modeling of global geogenic arsenic contamination in groundwater." *Environmental Science & Technology* **42**(10): 3669–3675.

Aulakh, M. S., M. P. S. Khurana et al. (2009). "Water pollution related to agricultural, industrial, and urban activities, and its effects on the food chain: Case studies from Punjab." *Journal of New Seeds* **10**(2): 112–137.

Baker, R. S., S. G. Nielsen et al. (2016). "How effective is thermal remediation of DNAPL source zones in reducing groundwater concentrations?" *Groundwater Monitoring & Remediation* **36**(1): 38–53.

Baldev, S. (1998). *Geoenvironmental Appraisal of Fatehgarh Sahib District, Punjab.* Kolkatta, India, Geodata Division, NR, GSI.

Balke, K.-D. and Y. Zhu (2008). "Natural water purification and water management by artificial groundwater recharge." *Journal of Zhejiang University Science B* **9**(3): 221–226.

Bamaga, O. A., S. F. Al-Sharif et al. (2016). "Improvement of urban water cycle and mitigation of groundwater table rise through advanced membrane desalination of shallow urban brackish groundwater of Jeddah basin." *Desalination and Water Treatment* **57**(1): 124–135.

Banerjee, A. (2015). "Groundwater fluoride contamination: A reappraisal." *Geoscience Frontiers* **6**(2): 277–284.

Barth, S. (1998). "Application of boron isotopes for tracing sources of anthropogenic contamination in groundwater." *Water Research* **32**(3): 685–690.

Böhlke, J.-K. (2002). "Groundwater recharge and agricultural contamination." *Hydrogeology Journal* **10**(1): 153–179.

Bouwer, H. (2002). "Artificial recharge of groundwater: Hydrogeology and engineering." *Hydrogeology Journal* **10**(1): 121–142.

Bukowski, J., G. Somers et al. (2001). "Agricultural contamination of groundwater as a possible risk factor for growth restriction or prematurity." *Journal of Occupational and Environmental Medicine* **43**(4): 377–383.

Central Ground Water Board (2014). *Water Quality Issues and Challenges in Punjab.* Faridabad, India, M. O. W. Resources.

Chakraborty, M., A. Mukherjee et al. (2015). "A review of groundwater arsenic in the Bengal Basin, Bangladesh and India: From source to sink." *Current Pollution Reports* **1**(4): 220–247.

Datta, P. S., D. L. Deb et al. (1997). "Assessment of groundwater contamination from fertilizers in the Delhi area based on 18O, NO3– and K+ composition." *Journal of Contaminant Hydrology* **27**(3): 249–262.

Davis, E. L. (1998). Steam injection for soil and aquifer remediation. *Ground Water Issue,* pp. 1–16.

Eulenstein, F., A. Saparov et al. (2016). Assessing and controlling land use impacts on groundwater quality. *Novel Methods for Monitoring and Managing Land and Water Resources in Siberia.* L. Mueller, K. A. Sheudshen and F. Eulenstein (Eds.), Cham, Switzerland, Springer International Publishing, pp. 635–665.

Farhadian, M., C. Vachelard et al. (2008). "*In situ* bioremediation of monoaromatic pollutants in groundwater: A review." *Bioresource Technology* **99**(13): 5296–5308.

Farooqi, A., H. Masuda et al. (2009). "Sources of arsenic and fluoride in highly contaminated soils causing groundwater contamination in Punjab, Pakistan." *Archives of Environmental Contamination and Toxicology* **56**(4): 693–706.

Finizio, A. and S. Villa (2002). "Environmental risk assessment for pesticides: A tool for decision making." *Environmental Impact Assessment Review* **22**(3): 235–248.

Foster, S. and H. Garduño (2013). "Groundwater-resource governance: Are governments and stakeholders responding to the challenge?" *Hydrogeology Journal* **21**(2): 317–320.

Foster, S., H. Garduno et al. (2010). Groundwater governance: Conceptual framework for assessment of provisions and needs. *GW MATE Strategic Overview Series.* Washington, DC, World Bank, p. 1.

Ghayoumian, J., M. Mohseni Saravi et al. (2007). "Application of GIS techniques to determine areas most suitable for artificial groundwater recharge in a coastal aquifer in southern Iran." *Journal of Asian Earth Sciences* 30(2): 364–374.

Giordano, M. (2009). "Global groundwater? Issues and solutions." *Annual Review of Environment and Resources* 34(1): 153–178.

Gogu, R. C. and A. Dassargues. (2000). "Current trends and future challenges in groundwater vulnerability assessment using overlay and index methods." *Environmental Geology* 39(6): 549–559.

Government of India, Ministry of MSME (2011). *Brief Industrial Profile of District Fatehgarh Sahib*. Ludhiana, India.

Gupta, J. and C. Vegelin. (2016). "Sustainable development goals and inclusive development." *International Environmental Agreements: Politics, Law and Economics* 16(3): 433–448.

Haase, D. (2009). "Effects of urbanisation on the water balance—A long-term trajectory." *Environmental Impact Assessment Review* 29(4): 211–219.

Hatzinger, P. B., M. C. Whittier et al. (2002). "*In-situ* and *ex-situ* bioremediation options for treating perchlorate in groundwater." *Remediation Journal* 12(2): 69–86.

Hellström, D., U. Jeppsson et al. (2000). "A framework for systems analysis of sustainable urban water management." *Environmental Impact Assessment Review* 20(3): 311–321.

Heron, G., K. Parker et al. (2009). "Thermal treatment of eight CVOC source zones to near nondetect concentrations." *Ground Water Monitoring & Remediation* 29(3): 56–65.

Howard, K. W. F. (2015). "Sustainable cities and the groundwater governance challenge." *Environmental Earth Sciences* 73(6): 2543–2554.

Karkra, R., P. Kumar et al. (2016). "Classification of heavy metal ions present in multi-frequency multi-electrode potable water data using evolutionary algorithm." *Applied Water Science* 6(5): 1–11.

Kosusko, M., M. E. Mullins et al. (1988). "Catalytic oxidation of groundwater stripping emissions." *Environmental Progress* 7(2): 136–142.

Krishna, A. K. and P. K. Govil (2004). "Heavy metal contamination of soil around Pali Industrial Area, Rajasthan, India." *Environmental Geology* 47(1): 38–44.

Kulkarni, H. and P. S. V. Shankar (2009). "Groundwater: Towards an aquifer management framework." *Economic and Political Weekly* 44(6): 13–17.

Kulkarni, H., M. Shah et al. (2015). "Shaping the contours of groundwater governance in India." *Journal of Hydrology: Regional Studies* 4(Part A): 172–192.

Kumar, P., B. K. Bansod et al. (2015). "Index-based groundwater vulnerability mapping models using hydrogeological settings: A critical evaluation." *Environmental Impact Assessment Review* 51: 38–49.

Kumar, P., P. K. Thakur et al. (2016). "Assessment of the effectiveness of DRASTIC in predicting the vulnerability of groundwater to contamination: A case study from Fatehgarh Sahib district in Punjab, India." *Environmental Earth Sciences* 75(10): 1–13.

Kumar, R., R. Kumar et al. (2016). "Role of soil physicochemical characteristics on the present state of arsenic and its adsorption in alluvial soils of two agri-intensive region of Bathinda, Punjab, India." *Journal of Soils and Sediments* 16(2): 605–620.

Kumar, T., A. K. Gautam et al. (2016). "Multi-criteria decision analysis for planning and management of groundwater resources in Balod district, India." *Environmental Earth Sciences* 75(8): 1–16.

Kuppusamy, S., T. Palanisami et al. (2016). *In-situ* remediation approaches for the management of contaminated sites: A comprehensive overview. *Reviews of Environmental Contamination and Toxicology*, Vol. 236. P. de Voogt (Ed.), Cham, Switzerland, Springer International Publishing, pp. 1–115.

Maleksaeidi, H., E. Karami et al. (2016). "Discovering and characterizing farm households' resilience under water scarcity." *Environment, Development and Sustainability* **18**(2): 499–525.

Marley, M. C., D. J. Hazebrouck et al. (1992). "The application of *in situ* air sparging as an innovative soils and ground water remediation technology." *Ground Water Monitoring & Remediation* **12**(2): 137–145.

McCray, J. E. and R. W. Falta (1996). "Defining the air sparging radius of influence for groundwater remediation." *Journal of Contaminant Hydrology* **24**(1): 25–52.

Minsker, B. S. and C. A. Shoemaker (1998). "Dynamic optimal control of *in-situ* bioremediation of ground water." *Journal of Water Resources Planning and Management* **124**(3): 149–161.

Mittal, S., G. Kaur et al. (2014). "Effects of environmental pesticides on the health of rural communities in the Malwa Region of Punjab, India: A review." *Human and Ecological Risk Assessment: An International Journal* **20**(2): 366–387.

Mukherji, A. and T. Shah (2005). "Groundwater socio-ecology and governance: A review of institutions and policies in selected countries." *Hydrogeology Journal* **13**(1): 328–345.

Nagarajan, R., N. Rajmohan et al. (2010). "Evaluation of groundwater quality and its suitability for drinking and agricultural use in Thanjavur city, Tamil Nadu, India." *Environmental Monitoring and Assessment* **171**(1): 289–308.

NGWA (1999). Principles of induced infiltration and artificial recharge. *Ground Water Hydrology for Water Well Contractors*. Westerville, OH, NGWA Press Publication.

Niroula, G. S. and G. B. Thapa (2005). "Impacts and causes of land fragmentation, and lessons learned from land consolidation in South Asia." *Land Use Policy* **22**(4): 358–372.

Pandey, R. (2016). Groundwater irrigation in Punjab: Some issues and a way forward. *Economic Transformation of a Developing Economy: The Experience of Punjab, India*. L. Singh and N. Singh (Eds.), Singapore, Springer, pp. 97–117.

Rabideau, A. J., J. M. Blayden et al. (1999). "Field performance of air-sparging system for removing TCE from groundwater." *Environmental Science & Technology* **33**(1): 157–162.

Ramakrishnaiah, C. R., C. Sadashivaiah et al. (2009). "Assessment of water quality index for the groundwater in Tumkur Taluk, Karnataka state, India." *E-Journal of Chemistry* **6**(2): 523–530.

Rao, S. M., K. Asha et al. (2013). "Influence of anthropogenic contamination on groundwater chemistry in Mulbagal town, Kolar District, India." *Geosciences Journal* **17**(1): 97–106.

Rasool, A., A. Farooqi et al. (2016). "Arsenic in groundwater and its health risk assessment in drinking water of Mailsi, Punjab, Pakistan." *Human and Ecological Risk Assessment: An International Journal* **22**(1): 187–202.

Saha, D. and S. Sahu (2016). "A decade of investigations on groundwater arsenic contamination in Middle Ganga Plain, India." *Environmental Geochemistry and Health* **38**(2): 315–337.

Saigal, S. K. (2007). *Ground Water Information Booklet Fatehgarh Sahib District, Punjab*. Chandigarh, India, Central Groundwater Board North Western Region.

Shah, T. (2005). "Groundwater and human development: Challenges and opportunities in livelihoods and environment." *Water Science and Technology* 51(8): 27–37.

Shah, T. (2014). *Groundwater Governance and Irrigated Agriculture.* Stockholm, Sweden, Global Water Partnership Technical Committee (TEC).

Shah, T., A. D. Roy et al. (2003). "Sustaining Asia's groundwater boom: An overview of issues and evidence." *Natural Resources Forum* 27(2): 130–141.

Sharma, C., A. Mahajan et al. (2016). "Fluoride and nitrate in groundwater of southwestern Punjab, India—occurrence, distribution and statistical analysis." *Desalination and Water Treatment* 57(9): 3928–3939.

Sikdar, P. K., P. Sahu et al. (2013). "Migration of arsenic in multi-aquifer system of southern Bengal Basin: Analysis via numerical modeling." *Environmental Earth Sciences* 70(4): 1863–1879.

Srinivasa Gowd, S., M. Ramakrishna Reddy et al. (2010). "Assessment of heavy metal contamination in soils at Jajmau (Kanpur) and Unnao industrial areas of the Ganga Plain, Uttar Pradesh, India." *Journal of Hazardous Materials* 174(1–3): 113–121.

Suthar, S., P. Bishnoi et al. (2009). "Nitrate contamination in groundwater of some rural areas of Rajasthan, India." *Journal of Hazardous Materials* 171(1–3): 189–199.

Tariq, M. I., S. Afzal et al. (2004). "Pesticides in shallow groundwater of Bahawalnagar, Muzafargarh, D.G. Khan and Rajan Pur districts of Punjab, Pakistan." *Environment International* 30(4): 471–479.

Tengberg, A. (2015). "World water week 2015." *Environment, Development and Sustainability* 17(6): 1247–1249.

Thakur, T., M. S. Rishi et al. (2016). "Elucidating hydrochemical properties of groundwater for drinking and agriculture in parts of Punjab, India." *Environmental Earth Sciences* 75(6): 1–15.

Tziritis, E., K. Skordas et al. (2016). "The use of hydrogeochemical analyses and multivariate statistics for the characterization of groundwater resources in a complex aquifer system. A case study in Amyros River basin, Thessaly, central Greece." *Environmental Earth Sciences* 75(4): 1–11.

UNESCO's International Hydrological Programme (UNESCO-IHP) (2015). Global Framework for Action to achieve the vision on Groundwater Governance.

Wezel, A. and S. Weizenegger (2016). "Rural agricultural regions and sustainable development: A case study of the Allgäu region in Germany." *Environment, Development and Sustainability* 18(3): 717–737.

White, D., D. J. Lapworth et al. (2016). "Hydrochemical profiles in urban groundwater systems: New insights into contaminant sources and pathways in the subsurface from legacy and emerging contaminants." *Science of the Total Environment* 562: 962–973.

Williams, A. E., L. J. Lund et al. (1998). "Natural and anthropogenic nitrate contamination of groundwater in a rural community, California." *Environmental Science & Technology* 32(1): 32–39.

Zektser, I. S., E. Y. Potapova et al. (2012). "Perspectives of artificial recharge of groundwater in southern European Russia." *Water Resources* 39(6): 672–684.

Zhu, X.-H., S.-S. Lyu et al. (2016). "Heavy metal contamination in the lacustrine sediment of a plateau lake: Influences of groundwater and anthropogenic pollution." *Environmental Earth Sciences* 75(2): 1–14.

Index

Printed in the United States
by Baker & Taylor Publisher Services